ESR Applications to Polymer Research

Nobel Symposium 22

ESR Applications to Polymer Research

Proceedings of the Twenty-Second Nobel Symposium
held June 20–22, 1972
at Södergarn, Lidingö, Sweden

Edited by

PER-OLOF KINELL

Professor of Physical Chemistry,
University of Umeå,
Umeå, Sweden

BENGT RÅNBY

Professor of Polymer Technology,
The Royal Institute of Technology,
Stockholm, Sweden

and VERA RUNNSTRÖM-REIO

ALMQVIST & WIKSELL
Stockholm

A Halsted Press Book

JOHN WILEY & SONS
New York - London - Sydney - Toronto

© 1973

Almqvist & Wiksell Förlag AB, Stockholm

Library of Congress Catalog Card Number 73-2847

Wiley ISBN 0-470-47770-9

Almqvist & Wiksell ISBN 91-20-03397-4

Cat/Sep
Chem

Printed in Sweden by

Almqvist & Wiksell Informationsindustri AB, Uppsala 1973

sd
5/29/73

Preface

The Twentysecond Nobel Symposium was held during June 20–22, 1972 at Södergarn, Lidingö, Stockholm. The theme of the symposium was *Electron Spin Resonance Applications to Polymer Research*. Since the initiation of the experimental method by Zavoisky in 1945, the use of the ESR technique has become quite widespread. It has been applied to many branches of chemistry where species having unpaired electron spins occur. The results obtained have in most cases been very informative and useful, and more so as the method has developed towards increased sensitivity and precision. It is a powerful method for free radical studies.

It has long been known that polymers can be formed by free radical mechanisms. The ESR technique offers unique possibilities to study the structure and reactions of these radicals. This is true not only for neutral radicals but also for positively and negatively charged species. Also degradation reactions of organic polymers, initiated with photons or high energy radiation, frequently involve free radical intermediates. Since the ESR method has been applied at an increasing rate to quite a variety of polymer problems in several laboratories, it was found appropriate to have a meeting where representatives from leading institutes in the field could get together and discuss experiences from their work and consider future prospects. A suggestion to the Nobel Foundation to arrange a Nobel Symposium covering the field of ESR applications in polymer research was met with great interest and approved. Accordingly, invitations were sent out to the most active and prominent scientists using the ESR technique for selected polymer problems. The invitations were met with a very positive response. The main fields to be covered were: initiation reactions including the use of high energy radiation, polymerization reactions, crosslinking reactions, photochemical degradation and oxidation, mechanical fracture of polymers, application of MO calculations, and molecular motions of polymers. The symposium also included a general discussion.

The papers presented and the discussion remarks made are collected in this volume. Its content gives examples of the use of the ESR technique in most established areas of polymer research and also hints towards new initiating systems, new polymer applications, and new methods to derive information from the spectra observed. Without doubt, the presentations given demonstrate that the ESR method is a most valuable tool in polymer research.

The location of the symposium at Södergarn holiday home afforded frequent informal discussions among the participants, even during a sightseeing boat trip in the Stockholm archipelago and at the symposium dinner.

Due to the devotion and skill of the session chairmen, the discussions were lively and fruitful. With the cooperation of the participants, it has been possible to summarize most of the discussion remarks to be included in the symposium volume.

Thanks are due to Svenska Handelsbanken for placing the facilities of the Södergarn holiday home at disposal for the symposium.

To Miss Kerstin Burman we wish to convey our gratitude for her devoted efforts to the organisation of the symposium.

The symposium was sponsored by the Nobel Foundation through grants from the Tricentennial Fund of the Bank of Sweden, which is gratefully acknowledged.

Stockholm, June 22, 1972

Bengt Rånby   *Per-Olof Kinell*

Contents

Initiation Reactions

Chemical and Photochemical Initiation

High Energy Initiation

Polymerization Reactions

Crosslinking Reactions

Photochemical Degradation and Oxidation

Mechanical Fracture of Polymers

Molecular Orbital Calculations

Molecular Motion in Polymers

Sponsors

The Nobel Foundation
The Tri-Centennial Fund of the Bank of Sweden

Nobel Foundation Symposium Committee

Ståhle, Nils K., Chairman, Executive Director of the Nobel Foundation
Ramel, Stig, Chairman, Executive Director of the Nobel Foundation (from July 1, 1972)
Hulthén, Lamek, Professor, Member of the Nobel Committee for Physics
Fredga, Arne, Professor, Chairman of the Nobel Committee for Chemistry
Gustafsson, Bengt, Professor, Secretary of the Nobel Committee for Medicine
Gierow, Karl Ragnar, Dr. Ph., Permanent Secretary of the Swedish Academy and Chairman of the Nobel Committee for Literature
Schou, August, Director of the Norwegian Nobel Institute (Peace)

Organizers of the Symposium

Bengt Rånby
Per-Olof Kinell
Vera Runnström-Reio

List of Participants

GÖRAN CANBÄCK, Department of Polymer Technology, The Royal Institute of Technology, S-100 44 Stockholm 70, Sweden

PETER CARSTENSEN, A/S Nordiske Kabel- og Traadfabriker, 7 La Cours Vej, DK-2000 Copenhagen F, Denmark

ADOLPHE CHAPIRO, Laboratoire de Chimie des Radiations, Laboratoires de Bellevue, 1, Place A. Briand, 92-Bellevue, France

ARTHUR CHARLESBY, Department of Physics, The Royal Military College of Science, Shrivenham, Swindon, Wilts SN6 8LA, UK

TOMAS GILLBRO, The Swedish Research Councils' Laboratory, Studsvik, Fack, S-611 01 Nyköping 1, Sweden

PÉTER HEDVIG, Research Institute for the Plastics Industry, Hungária Körût 114, Budapest XIV, Hungary

THORMOD HENRIKSEN, Department of Biophysics, Institute of Physics, University of Oslo, P.O. Box 1048, Blindern, Oslo 3, Norway

K. J. IVIN, Department of Chemistry, The Queen's University of Belfast, Belfast BT9 5AG, Northern Ireland

ZENZI IZUMI, Department of Polymer Technology, The Royal Institute of Technology, S-100 44 Stockholm 70, Sweden

HISATSUGU KASHIWABARA, Nagoya Institute of Technology, Gokiso, Showa-Ku, Nagoya 466, Japan

IKRAM KHOKHAR, Department of Polymer Technology, The Royal Institute of Technology, S-100 44 Stockholm 70, Sweden

PER-OLOF KINELL, Department of Physical Chemistry, University of Umeå, S-901 87 Umeå, Sweden

PEKKA LEHMUS, Neste Oy, Tutkimuskeskus, Kulloo, Finland

J. JOHAN LINDBERG, Department of Wood and Polymer Chemistry, University of Helsinki, Malminkatu 20, Helsinki 10, Finland

ANDERS LUND, The Swedish Research Councils' Laboratory, Studsvik, Fack, S-611 01 Nyköping 1, Sweden

SWARAJ PAUL, Department of Polymer Technology, The Royal Institute of Technology, S-100 44 Stockholm 70, Sweden

ANTON PETERLIN, Camille Dreyfus Laboratory, Research Triangle Institute, P.O. Box 12194, Research Triangle Park, N. C. 27709, USA

JAN F. RABEK, Department of Polymer Technology, The Royal Institute of Technology, S-100 44 Stockholm 70, Sweden

VERA RUNNSTRÖM-REIO, Karolinska Institutet, S-104 01 Stockholm 60, Sweden

BENGT RÅNBY, Department of Polymer Technology, The Royal Institute of Technology, S-100 44 Stockholm 70, Sweden

DON R. SMITH, Physical Chemistry Branch, Atomic Energy of Canada Ltd., Chalk River Nuclear Laboratories, Chalk River, Ontario, Canada

PETER SMITH, Paul M. Gross Chemical Laboratory, Duke University, Durham, N.C. 27706, USA

JUNKICHI SOHMA, Department of Materials Science, Faculty of Engineering, Hokkaido University, Sapporo 060, Japan

MICHAEL SZWARC, State University College of Forestry at Syracuse University, Syracuse, N.Y. 13210, USA

KOZO TSUJI, Sumitomo Chemical Co., Ltd., Central Research Laboratory, 40 Tsukahara, Takatsuki City, Osaka, Japan

FFRANCON WILLIAMS, Department of Chemistry, The University of Tennessee, Knoxville, Tenn. 37916, USA

HIROSHI YOSHIDA, Faculty of Engineering, Hokkaido University, Sapporo 060, Japan

Participants invited to the last section meeting

OVE EDLUND, The Swedish Research Councils' Laboratory, Studsvik, Fack, S-611 01 Nyköping 1, Sweden

GÖRAN ERIKSSON, Department of Biophysics, University of Stockholm, c/o Karolinska Institutet, Solnavägen 1, 104 01 Stockholm 60, Sweden

ERIK FORSLIND, Department of Physical Chemistry, The Royal Institute of Technology, S-100 44 Stockholm 70, Sweden

ARNE FREDGA, Department of Organic Chemistry, University of Uppsala, 751 05 Uppsala, Sweden

TORBJÖRN REITBERGER, Department of Nuclear Chemistry, The Royal Institute of Technology, S-100 44 Stockholm 70, Sweden

S. E. SWANSON, Department of Physical Chemistry, University of Göteborg, Gibraltargatan 5A, 402 20 Göteborg, Sweden

List of Chairmen

Working group: Initiation reactions
   Chairman: Michael Szwarc

Working group: High energy initiation
   Chairman: Arthur Charlesby

Working group: Polymerization reactions
   Chairmen: Péter Hedvig
                J. Sohma

Working group: Crosslinking reactions
   Chairman: Adolphe Chapiro

Working group: Photochemical degradation and oxidation
   Chairman: H. J. Ivin
                Anton Peterlin

Working group: Mechanical fracture of polymers
   Chairman: Ffrancon Williams

Working group: Molecular motion in polymers
   Chairman: Bengt Rånby

INITIATION REACTIONS

Chemical and Photochemical Initiation

ESR Spectra of Radicals Formed in the Reaction between *t*-Butyl Hydroperoxide and Sulphur Dioxide in the Presence of Unsaturated Compounds

By B. D. Flockhart, K. J. Ivin, R. C. Pink and B. D. Sharma[1]

Department of Chemistry, The Queen's University of Belfast, UK

Over the past ten years many workers (Dixon & Norman, 1963; Yoshida & Rånby, 1967; Fischer, 1965) have shown that radical intermediates can be detected in flowing reacting systems in good yield (10^{-5} mol l^{-1}) by ESR spectroscopy. Vinyl monomers M, when added to redox systems in which radicals R· are being generated, frequently show evidence of radicals of the type RM·. Such studies provide a wealth of information concerning the structure, spin density and conformation of these radicals; also concerning the reactivity of the monomers and the presence of other types of radical. An important feature of the spectra, which until recently has received comparatively little attention, is their frequent lack of symmetry, stemming from a non-Boltzmann population of the energy levels.

The initiator system, *t*-butyl hydroperoxide + sulphur dioxide, has been comparatively little used until recently (Mazzolini et al., 1970) as an initiator for vinyl polymerization. Here we summarize our results from ESR studies on this initiator system in methanol, with 28 additives. Brief accounts of the primary reaction, which produces $HO\check{S}O_2$ and t-$BuO\check{S}O_2$, and the effects of some additives have been published (Flockhart et al., 1971 a, b). The experimental arrangement was similar to that of Dixon & Norman (1963).

The Primary Reaction

Conductance measurements at low temperature ($-76°$ to $-46°C$) showed that the reaction in methanol was first order with respect to each reactant, with a rate constant $k = 2.2 \times 10^7 \exp(-29.8 \text{ kJ}/RT) l \, mol^{-1} \, s^{-1}$. Extrapolation to 20°C, the average temperature of the ESR experiments, gives $k = 110 \, l \, mol^{-1} \, s^{-1}$. For initial reactant concentrations (after mixing) of 0.1 M the half-life for

[1] Present address: Department of Radiology, University of Chicago, Chicago, Ill. 60637, USA.

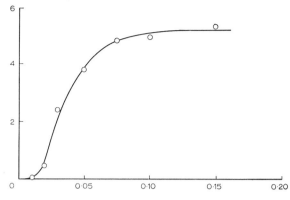

Fig. 1. *Abscissa:* $[A_0]$/mol l^{-1}; *ordinate:* $10^5 \times$ radical concentration/mol l^{-1}.
   Variation of observed radical concentration with initial (equal) reactant concentrations $[A_0]$, in methanol at 20°C. Carbon samples of known radical content were used as standards.

consumption of reactants is therefore 0.090 s which is slightly longer than the time (0.080 s) between mixing and observation (centre of the ESR cavity). The reactant concentrations have thus fallen to about 0.05 M at the point of observation.

   As a check on the rate constant and as a means of estimating the radical lifetime we determined the concentration of radicals as a function of initial (equal) reactant concentrations, at constant flow rate. Fig. 1 shows the observed behaviour. This may be interpreted in terms of the scheme ($A = t$-BuOOH, $B = SO_2$):

$$A + B \xrightarrow{k_1} 2R \quad 2R \xrightarrow{k_t} \text{products}$$

The radicals R may consist of more than one type and all radicals are assumed to disappear by mutual interaction with the same rate constant k_t. With equal initial concentrations $[A_0]$ it is readily shown that after a reaction time t, the stationary radical concentration $[R]$ is given by $(k_1/k_t)^{\frac{1}{2}}[A_0]$ at low $[A_0]$, and by $1/(k_1 k_t)^{\frac{1}{2}} t$ at high $[A_0]$. The curve shown in Fig. 1 is of this form except for the anomaly at low $[A_0]$, which is attributed to inefficient mixing. From the initial slope and plateau, and with $t = 0.083$ s, we find $k_1 = 240$ l mol^{-1} s^{-1} and $k_t = 2.4 \times 10^8$ l mol^{-1} s^{-1}. k_1 is somewhat higher than that estimated from low temperature kinetics (see above) but the agreement can be considered satisfactory in view of the approximations and assumptions required in arriving at a value from ESR measurements. Certainly we may conclude that the reaction proceeds largely by a radical mechanism in methanol as solvent.

   The lifetime of the radicals at the point of observation by ESR, for $[A_0] = [B_0] = 0.1$ M, is given by $1/k_t[R] = 0.8 \times 10^{-4}$ s. Since the time for the solution to flow through the sensitive part of the cavity is about 10^{-2}s we arrive at the important conclusion that the observed spectra are those of radicals which are born and which react and die within the cavity. Virtually none of the radicals generated between the mixing point and the upper end of the cavity survive long enough to be detected. This situation is probably still the same when monomer is present. Czapski (1971) gives an interesting discussion of this sometimes misunderstood point. The value of k_t is about 100 times smaller than the collision number, which seems reasonable for polar radicals of the type ROŠO₂.

Effects of Added Unsaturated Compounds

The characteristics of the spectra which appear in the presence of unsaturated compounds are summarized in Table 1 and a few typical spectra are shown in Fig. 2.

Types of Radical Observed

Most of the spectra are attributable to monomer radicals of structure RCH₂ĊHX, RCH₂ĊXY, or RCHXĊHY. In some cases there is a second, weaker, set of lines corresponding to a radical with slightly different coupling constants. For both styrene and vinyl acetate the appearance of the stronger set of lines correlates with the disappearance of HOŠO₂ while the weaker set correlates with the disappearance of t-BuOŠO₂, which is also the more reactive radical. The two sets of lines are thus attributed to HOSO₂M· and t-BuOSO₂M· (or possibly t-BuOM·). Radicals of the type HOM·, whose coupling constants are well known (Corvaja et al., 1965; Fischer, 1964; Griffiths et al., 1967; Yoshida & Rånby, 1967), were not detected.

Norman & Storey (1971), using the reaction HȮ + SO₃²⁻ → HO⁻ + SO₃⁻ in aqueous solution to generate the sulphite radical-ion, with subsequent addition to monomers, have observed spectra similar to ours (hfcc in gauss shown in brackets): acrylic acid (20.2, 15.6), acrylonitrile (20.0, 14.3, 3.4), vinyl acetate (20.2, 12.8, 1.3), maleic acid (20.2, 7.65), crotonic acid (20.0, 6.5, 0.4). With our initiating system and these monomers in *water* we observed the same hfcc as Norman & Storey. The solvent effect on the β-coupling constants for the acids probably stems from the different degrees of ionization of the SO₃H group in the two solvents. Two other points of difference were noted: Norman & Storey observed only RM· with allyl alcohol whereas we observed only RMŠO₂; and they did not observe reaction with acrylamide or methyl acrylate. The spectrum attributed to RM⁻ₙ with acrylic acid has been previously observed with R = HO, H₂N, H₃C in aqueous medium (20.4, 23.0 G) (Corvaja et al., 1965; Fischer et al., 1967). Only in the case of cyclopentene is there evidence for radicals formed by hydrogen abstraction; the 3-cyclopentenyl

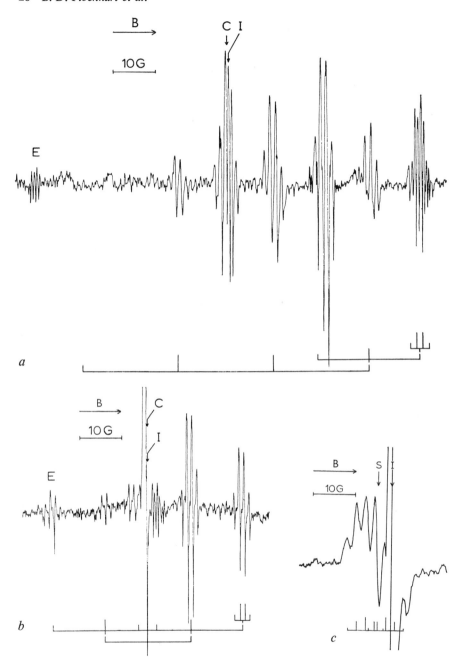

Fig. 2. ESR spectra for $[A_0] = [B_0] = 0.2$ mol 1^{-1} and $[M_0]$ as indicated: (*a*) vinyl acetate 0.4 M; (*b*) methyl methacrylate 0.1 M; (*c*) hex-1-ene 0.5 M; I, primary radical signal; C, centre of RM· spectrum; S, centre of $RMSO_2$ spectrum; E, emission lines.

radical has been previously identified with hfcc = 21.6, 14.2, 3.1 G (Gee & Wan, 1971). No abstraction was detected from saturated compounds such as isopropyl acetate.

All compounds capable of forming $1:1$ copolymers with SO_2 (hex-1-ene, allyl alcohol, isobutene, cyclopentene, cyclohexene, cycloheptene), showed evidence for the presence of $R M \check{S} O_2$. These radicals are easy to identify by their high g factors and low hfcc. Only with isobutene was there also a spectrum due to $RM\cdot$ indicating that the conditions were such that the equilibrium $RM\cdot + SO_2 \rightleftharpoons RM\check{S}O_2$ was reasonably balanced, as may be expected from ceiling temperature data (Cook et al., 1965).

Monomer Reactivities, Spin Densities and Dihedral Angles

We have previously given some account of these aspects of our work (Flockhart et al., 1971 a, b) and here we can only indicate the main conclusions. The reactivities of seven monomers studied lie in the following order: methacrylonitrile > acrylonitrile $\approx$ styrene > itaconic acid $\approx$ methacrylic acid $\gg$ vinyl acetate $\approx$ maleic acid. For methacrylonitrile a concentration of 0.02 M monomer was sufficient to reduce the primary radical concentration to a third of its original value; for maleic acid more than 0.5 M was required to produce the same effect. No correlation was found between these results and the relative reactivity of these monomers towards polymer radicals of various kinds.

Spin densities at the terminal carbon, ϱ_α, calculated from the hfcc in the usual way, ranged from 0.85 for the acrylic acid radical to 0.61 for the styrene radical. $\Delta(X_i)$ values in the empirical relationship $\varrho_\alpha = \prod_{i=1}^{3} (1 - \Delta(X_i))$ were estimated as 0.085 for $ROSO_2CH_2$ and 0.330 for C_6H_5.

The β-coupling constants of the radicals $RM\cdot$ $(R = R'OSO_2)$ are considerably smaller than those for $R = HO$, H_2N, H_3C, or $HOCH_2$ (Fischer et al., 1967) but similar to those for $R = R'S$ (Krusic & Kochi, 1971; Kawamura et al., 1971). This implies smaller dihedral angles (the angle between the axis of the $2 p_z$ orbital and the projection of the C^β–R bond in the plane passing through this axis and perpendicular to the C^β–C^α bond) for $R = R'OSO_2$ or $R'S$ in the minimum energy conformations. This may be interpreted partly as a steric effect, and partly in terms of some specific interaction between the R group and the electron density on the α carbon.

Asymmetry of the Spectra

The magnitude of this effect is indicated in Table 1. It is greater, the wider the separation of the lines and for this reason we have compared the asymmetry for different monomer radicals at approximately constant field difference. With acrylic acid the degree of asymmetry (defined in Table 1) declined from 1.57 to 1.30 as [M] increased from 0.05 M to 0.5 M. The asymmetry was least

Table 1. Radicals obtained from t-Butyl Hydroperoxide and Sulphur Dioxide in the Presence of Additives (Solvent, Methanol)

Additive M	Radical	Type of spectrum A × B × C	Hfcc (G) A	B	C	g	Degree of asymmetry^{a,c}	Notes
CH₂ = CHX								
Acrylic acid	$RCH_2^B\dot{C}H^ACOOH$	2 × 3	20.4	17.2		2.0037	1.55/20.4G/0.1 M	Two sets of additional lines which disappear at [M] > 0.5
	$RM_n^{\cdot}$	2 × 3	20.4	23.2		2.0036		
Methyl acrylate	$RCH_2^B\dot{C}H^ACOOCH_3^C$	2 × 3 × 4	20.4	16.7	1.5	2.0039	1.32/21.9G/0.05 M	
Ethyl acrylate	$RCH_2^B\dot{C}H^ACOOCH_2^CCH_3$	2 × 3 × 3	20.4	16.8	1.4	2.0038	1.20/20.4G/0.05 M	
Acrylamide	$RCH_2^B\dot{C}H^ACON^CH_2$	2 × 3 × 5	20.0	16.8	2	2.0033	~2/20G/0.2 M	$a_N \approx a_H$ for NH₂ group
Methyl vinyl ketone	$RCH_2^B\dot{C}H^ACOCH_3^C$	2 × 3 × 4	19.1	15.9	1.0	2.0045	1.21/19.1G/0.02 M	Some additional lines
Acrylonitrile	$RCH_2^B\dot{C}H^ACN^C$	2 × 3 × 3	20.4	14.5	3.4	2.0033	1.43/20.4G/0.1 M	
Vinyl acetate	$RCH_2^B\dot{C}H^AOCOCH_3^C$	2 × 3 × 4	20.4	12.9	1.4	2.0033	1.55^b/21.8G/0.4 M	Some additional lines. Fig. 2 a
Vinyl propionate	$RCH_2^B\dot{C}H^AOCOCH_2^CCH_3$	2 × 3 × 3	19.5	13.2	1.5	2.0033	1.31/19.5G/0.2 M	Some additional lines
Vinyl isobutyrate	$RCH_2^B\dot{C}H^AOCOCH^C(CH_3)_2$	2 × 3 × 2	19.6	12.0	1.0	2.0033	2.7^b/18.6G/0.2 M	Some additional lines
Vinyl chloride	$RCH_2^B\dot{C}H^ACl^C$	2 × 3 × 4	20.8	14.8	2.8	2.0062	1.0/58.8G/saturated	Symmetric spectrum
Styrene	$RCH_2^B\dot{C}H^AC_6H_5$	2 × 3 × 3 ×3 ×2	15.32	12.12	5.00 (o) 1.76 (m) 5.92 (p) ^{d}	2.0029	1.65/21G/0.3 M	Main spectrum attributed to HOSO₂M·
			^{d}	10.6	^{d}			Attributed to t-BuOM· or t-BuOSO₂M·
Hex-1-ene	$RCH_2^BCH^A(CH_2^BPr^n)\dot{S}O_2$	2 × 5	4.7	2.1		2.0055	~1/9.5G/0.5 M	Overlap causes spectrum to show only 7 lines. Fig. 2 c
Allyl alcohol	$RCH_2^BCH^A(CH_2^BOH)\dot{S}O_2$	2 × 5	4.8	1.6		2.0055	>1/11.2G/0.8 M	Overlap causes spectrum to show only 8 lines
CH₂ = CXY								
Methacrylic acid	$RCH_2^B\dot{C}(C(CH_3^A)COOH$	4 × 3	22.6	12.0		2.0035	1.59^b/22.6G/0.3 M	Additional 2 × 2 spectrum (hfcc 23.6, 10.0) seen at low concentrations of monomer only
Methyl methacrylate	$RCH_2^B\dot{C}(CH_3^A)COOCH_3^C$	4 × 3 × 4	22.4	12.0	1.4	2.0037	1.55^b/22.4G/0.1 M	Fig. 2 b
	$RCH_2^B\dot{C}(CH_3^A)COOCH_3^C$	4 × 3 × 4	22.4	11.3	1.4	2.0037		
Ethyl methacrylate	$RCH_2^B\dot{C}(CH_3^A)COOCH_2^CCH_3$	4 × 3 × 3	22.4	12.0	1.5	2.0036	1.68^b/22.4G/0.05 M	Some additional lines

Monomer / radical source	Radical	Multiplicity	hfcc (G)	g	Rel. int. / field / monomer conc.[a]	Remarks
Methacrylonitrile	$RCH_2\dot{C}(CH_3)CN$	4 × 3 × 5 / 4 × 3 × 3	21.2, 11.2, 3.2 / 21.8, 10.8, 3.2	2.0031	1.30/21.2/… M	Weak set of lines
Itaconic acid	$RCH_2^B\dot{C}(CH_2^A COOH)COOH$	3 × 3	13.6, 12.5	2.0036	1.32/25.0G/0.8 M	Spectrum approximates to a quintet
Dimethyl itaconate	$RCH_2^B\dot{C}(CH_2^A COOCH_3)COOCH_3^C$	3 × 3 × 4	13.6, 13.2, 1.4	2.0036	1.25/27G/0.07 M	Spectrum approximates to five quartets
Isopropenyl acetate	$RCH_2^B\dot{C}(CH_3^A)OOCCH_3^C$	4 × 3 × 4	22.8, 11.6, 0.5	2.0033	~1.4/22.8G/0.2 M	Spectrum approximates to nine quartets
α-Methylstyrene	$RCH_2^B\dot{C}(CH_3^A)C_6H_5^C$	4 × 3 × 3 × 3 × 2	16.52, 9.40, 4.80 (o), 1.64 (m), 5.40 (p)	2.0029	asymmetric/0.2 M	
Isobutene	$RCH_2^B\dot{C}(CH_3^A)_2$	7 × 3	23.0, 11.5	2.0033	asymmetric/sat. solution	Weak signal, fine structure not resolved; *cf.* $g = 2.0055$ for $CH_3\dot{S}O_2$
	$RCH_2C(CH_3)_2\dot{S}O_2$	1		2.005		
CHX = CHY						
Maleic acid	$RCH^B(COOH)\dot{C}H^A COOH$	2 × 2	20.4, 6.4	2.0037	1.43/20.4G/0.05 M	Same spectrum with fumaric acid
Crotonic acid	$RCH^B(CH_3^C)\dot{C}H^D CH^A COOH$	2 × 2 × 4	20.4, 6.0, 1.0	2.0033	asymmetric/0.1 M	Weak spectrum
Cyclopentene	$\overline{CH_2CH_2CH^A CH^B} = CH^C\dot{C}H^B$	5 × 3 × 2	22.5, 14.4, 2.8	2.0030	asymmetric/0.1 M	Fine structure overlaid by spectrum of other radical
	$RM\dot{S}O_2$			2.0055		
Cyclohexene	$RM\dot{S}O_2$	3	3.6	2.0055	1.0/7.2G/0.2 M	Only two protons interact
Cycloheptene	$RM\dot{S}O_2$	2 × 2 × 3	5.4, 4.2, 2.4	2.0054	1.0/14.4G/0.2 M	All α and β protons interact; contrast cyclohexene
Primary radicals for comparison						
	$HO\dot{S}O_2$	1		2.0033		Not observed in aqueous medium
	$(CH_3)_3CO\dot{S}O_2$	10	0.28	2.0034		Not observed in aqueous medium
	$CH_3\dot{S}O_2$	4	0.9	2.0055		Only observed in aqueous medium

[a] Defined as observed high field intensity line, divided by average of high and corresponding low field intensities, separated by field indicated after first oblique stroke; concentration of monomer indicated after second oblique stroke. Accuracy in hfcc is ±0.2G and in g ±0.002.

[b] Lowest field lines of main spectrum appear in emission.

[c] Concentrations of monomer in Table 1 and in Fig. 2 are before mixing. Actual concentrations are about half these values.

[d] As for main spectrum.

with vinyl chloride and greatest with vinyl isobutyrate; emission lines were seen at low field in five cases.

This effect has its parallel in the anomalous NMR spectra for the *products* of radical reactions immediately after formation (CIDNP). Both effects are due to non-Boltzmann populations of the various energy levels and may be explained in terms of the radical-pair theory developed by Closs (1971), Kaptein & Oosterhoff (1969) and Adrian (1971). The essential feature of the theory is that a given radical pair has a proportion of singlet character and a proportion of triplet character, the proportions fluctuating with time in a magnetic field at a rate (approx. 10^8 Hz) which is dependent on the nuclear spin states of those nuclei which interact with the unpaired electrons. The rate of mutual destruction of the radicals is assumed to be dependent on the proportion of singlet character. For radicals which are generated independently of one another, as in the present case (leading to "uncorrelated" radical pairs), this results in initial overpopulation of the upper nuclear hyperfine levels for each electron state and hence to emission or reduced absorption in the low field lines and to enhanced absorption in the high field lines. At the same time relaxation processes tend to restore the Boltzmann population within a period of about 10^{-6}–10^{-4} s^{-1}. The reason that the effect is seen at all is that the radical lifetimes are in this region (see above). The variation of the observed effect with radical structure may thus be understood in terms of variable radical lifetime, which depends on the radical concentration and the effective value of k_t, and variable relaxation time, which depends on the radical structure and its environment. It is notable that the effect is smallest for radicals with high g factor where spin-orbit coupling is expected to reduced the relaxation time.

References

Adrian, F. J., J. Chem. Phys., *54*, 3918 (1971).

Closs, G. L., XXIIIrd Int. Congress of Pure & Applied Chem., Boston, Special lectures, *4*, 19 (1971).

Cook, R. E., Ivin, K. J. & O'Donnell, J. H., Trans. Faraday Soc., *61*, 1887 (1965).

Corvaja, H., Fischer, H. & Giacometti, G., Z. phys. Chem., (Frankfurt), *45*, 1 (1965).

Czapski, G., J. Phys. Chem., *75*, 2957 (1971).

Dixon, W. T. & Norman, R. O. C., J. Chem. Soc., 3119 (1963).

Fischer, H., Z. Naturforsch., *19a*, 866 (1964).

Fischer, H., Z. Naturforsch., *20a*, 488 (1965).

Fischer, H. & Giacometti, G., J. Polymer Sci., *C16*, 2763 (1967).

Flockhart, B. D., Ivin, K. J., Pink, R. C. & Sharma, B. D., J. Chem. Soc., D, 339 (1971*a*).

Flockhart, B. D., Ivin, K. J., Pink, R. C. & Sharma, B. D., XXIIIrd Int. Congress of Pure & Appl. Chem., Boston (1971*b*), preprints p. 319.

Gee, D. R. & Wan, J. K. S., Canad. J. Chem., *49*, 20 (1971).

Griffiths, W. E., Longster, G. F., Myatt, J. & Todd, P. F., J. Chem. Soc., B, 530 (1967).

Kaptein, R. & Oosterhoff, J. L., Chem. Phys. Letters, *4*, 195 (1969).

Kawamura, T., Ushio, M., Fujimoto, T. & Yonezawa, T., J. Amer. Chem. Soc., *93*, 908 (1971).

Krusic, P. J. & Kochi, J. K., J. Amer. Chem. Soc., *93*, 846 (1971).

Mazzolini, C., Patron, L., Moretti, A. & Campanelli, M., Ind. Eng. Chem. (Prod. Res. & Development), *9*, 504 (1970).

Norman, R. O. C. & Storey, P. M., J. Chem. Soc. B, 1009 (1971).

Yoshida, H. & Rånby, B., J. Polymer Sci., *C16*, 1333 (1967).

Discussion

P. Smith

As you have correctly pointed out, the spectra observed for transient addition radicals in the liquid phase have been known for some time to show asymmetry. Your comment that for a given radical, this asymmetry depends on its kinetic lifetime and its relaxation characteristics relates to our results to be given in my paper, pp 29–36. In our work we find that the $(CH_3)_2\dot{C}CN$ radical does not show this effect to a significant extent, whereas the addition radicals from the alkyl methacrylates do. In our work we have argued that the kinetic lifetime of a $(CH_3)_2\dot{C}CN$ radical and of an addition radical may not be very different. As I interpret your comment, it is possible that the lack of asymmetry in the spectrum from the $(CH_3)_2\dot{C}CN$ radical and clear the asymmetry in the spectra from the addition radicals may well be a result of a difference in relaxation characteristics?

Ivin

The degree of asymmetry will depend not only on the relaxation time of the radicals but also on their reaction half-life $t_\frac{1}{2}$. If $\tau \gg t_\frac{1}{2}$ one can expect to see a very asymmetric spectrum, whereas if $\tau \ll t_\frac{1}{2}$ the spectrum will be symmetrical.

Changing from $(CH_3)_2\dot{C}CN$ to $(CH_3)_2C(CN)M\cdot$ evidently increases $\tau/t_\frac{1}{2}$ in your case. Writing the first radical as R_1 and the second as R_2 we can consider two *limiting* cases, assuming the same rate of initiation I, and mutual destruction of radicals in each case:

(1) No monomer present

$$I = (k_t)_{R_1}[R_1]^2, \quad [R_1] = (I/(k_t)_{R_1})^\frac{1}{2}, \quad (t_\frac{1}{2})_{R_1} = (I(k_t)_{R_1})^{-\frac{1}{2}}$$

(2) Monomer present

$$I = (k_t)_{R_2}[R_2]^2, \quad [R_2] = (I/(k_t)_{R_2})^\frac{1}{2}, \quad (t_\frac{1}{2})_{R_2} = (I(k_t)_{R_2})^{-\frac{1}{2}}$$

$t_\frac{1}{2}$ is inversely proportional to the square root of the termination rate constant

k_t so that the appearance of asymmetry in the spectrum of R_2 could result from a higher k_t, i.e. $(k_t)_{R_2} > (k_t)_{R_1}$. In so far as cross-termination constants are generally higher than the termination constants of the individual radicals, one may expect that the degree of asymmetry of the product radical will fall as the monomer concentration is increased. This is the of the term "effective k_t," in our paper. As mentioned in our paper we observed a decline in the degree of asymmetry of the acrylic acid radical as the monomer concentration was increased.

Hedvig

With respect to the interpretation of the negative population of the energy levels is that absolutely necessary to assume radical pair formation? One could describe the state of a system containing randomly distributed radicals by using combined singlet-triplet eigenfunctions. This would mean that for considering population of the radical levels collective states should be considered instead of isolated molecular states.

Ivin

The maximum concentration of radical (Fig. 1) is of the order of 5×10^{-5} M, which means that on average the radicals are separated from each other by about 100 solvent molecules, or 50 nm, assuming that they are randomly disposed in the medium. This is almost certainly the case since the monomer concentrations were relatively low (0.1 M) and the primary radicals have to diffuse a certain distance before meeting and reacting with a monomer molecule. Also the asymmetry is always of the same type, namely enhanced absorption at high field and emission, or reduced absorption at low field which is consistent with the eventual reaction of "uncorrelated" radicals. It does not seem likely that collective states would be important except for an extremely small fraction of the radicals.

Henriksen

It is very difficult for me to realize that sulfur radicals should have a *g* value of the same order of magnitude (2.02) as that found for ordinary hydrocarbon radicals. Mostly the sulfur radicals are observed with an average *g* value of about 2.005. It is reasonable to assume that the spin density in your radicals are localized on other atoms with a smaller spin-orbit coupling constant.

Ivin

As indicated in Table 1, radicals of the type $R_1R_2R_3C\dot{S}O_2$, e.g. $CH_3\dot{S}O_2$, have *g* factors of 2.0055; the radical-ion SO_2^- is also known to have a high *g* factor. I agree that it is difficult to understand why the radicals $R_1R_2R_3O\dot{S}O_2$, $HO\dot{S}O_2$,

SO$_3^-$ do not have similar high g factors instead of values close to the free-spin values. One can certainly write down resonance forms of the type RSȮ but it is

$$\overset{\parallel}{O}$$

not easy to see why these should be more important in the latter radicals. I would suggest that there must be other reasons for the quenching of the spin-orbit interaction in the latter radicals. I do not think any explanation of this difference has yet been given.

Sohma

Do you think that the emission spectrum reported by you can be explained by the model similar to the Closs model for CIDNP even for the case of the ESR, especially ESR spectrum from the system having many coupled hydrogens? The spectrum which shows the emission should have the enhanced absorption in the other wing. Tell me please the distribution of the peak intensity from the emission side to the enhanced absorption side. Do you think this distribution can be explained by the model you proposed?

Ivin

Adrian (J. Chem. Phys., *54*, 3918 (1971)) has shown how the Closs model can be used for the quantitative interpretation of asymmetric ESR spectra observed in flowing systems. I have no doubt that the model can be used to interpret our spectra. The degree of asymmetry is a smoothly increasing function of the field difference between corresponding pairs of lines, as required by the theory.

Yoshida

I would like to make a comment about our experimental observation of an ESR spectrum of a neutral semiquinone radical observed during photolysis of 1,4-naphthoquinone in ethanol. It was found as emission at lower field part. The recovery time from emission to absorption was found to be about 10^{-1} s, much longer than the paramagnetic relaxation times, T_1 and T_2, about 10^{-6} s. This may give an example of emission spectra which cannot be interpreted by any theory so far reported.

ESR Spectroscopic Studies of the Photo-Initiated Polymerization of Alkyl Methacrylates in the Liquid Phase

By Peter Smith

Department of Chemistry, Paul M. Gross Chemical Laboratory, Duke University, Durham, N. C. 27706, USA

Introduction

There have been many ESR studies of radical addition to substituted ethylenes in the liquid phase (Smith & Wood, 1967; Fischer, 1968; Takakura & Rånby, 1969, 1970; Norman & Storey, 1971; Flockhart et al., 1971; Kawamura et al., 1971; Bullock, Burnett & Kerr, 1971 a; Kochi & Krusic, 1970; Doi & Rånby, 1971; Hefter, Hecht & Hammond, 1972; Neta, Hoffman & Simic, 1972). Most of these studies have been carried out with the use of thermal-redox radical-generating methods within rapid-mixing, continuous-flow systems. The nature of such methods so far found successful has limited the kind of solvent used, water being the usual choice, and hence also the type of substrate and the range of reaction conditions studied. A second limitation of these thermal-redox methods has stemmed from the necessary presence in the reaction mixture of the components of the redox system; for commonly these components may be the source of secondary reactions liable to complicate the overall kinetic analysis of the events taking place and, at the same time, may reduce the spectroscopic resolution attainable (Smith & Wood, 1967; Czapski, 1971).

In principle, these limitations may be circumvented more easily if the primary radicals are generated photolytically. Moreover, photolytic radical-generating methods have the added practical and theoretical advantages associated with a single-stream or a static-sample operation. However, few radical-addition reactions involving simple aliphatic species have been reported with the use of photolytic radical-generating methods (e.g., Hefter & Fischer, 1969; Kochi & Krusic, 1970; Hefter, Hecht & Hammond, 1972).

The present report concerns the use of a photolytic continuous-flow method at about room temperature to successfully observe radicals resulting from the addition of a photolytically generated primary radical to each of four methacrylic esters. The primary radical was the 2-cyano-2-propyl radical $(CH_3)_2\dot{C}CN$, R·, produced by the photodecomposition of 2,2'-azo-*bis*-isobutyronitrile, $(CH_3)_2C(CN)N:NC(CN)(CH_3)_2$, RN:NR, a well known photoinitiator for studies of radical polymerization (Oster & Yang, 1968). The methacrylic esters were of general formula, $CH_2:C(COOY)CH_3$ where Y was methyl, ethyl, *i*-butyl, and *n*-butyl, being denoted MMA, EMA, IBMA, and NBMA, respectively. For each ester, two ESR spectra were observed, one from R· and the other from addition radicals of general formula

$$R-[-CH_2-C(COOY)CH_3-]_n-CH_2-\dot{C}(COOY)CH_3, \tag{1}$$

with $n \geqslant 0$. The ESR spectrum of R· is well characterized already and our results for this radical are in general agreement with previous reports (e.g., Brumby, 1970). The coupling constants of the addition radicals are similar to those of the few comparable species successfully characterized with the use of thermal-redox radical-generating methods (Fischer, 1968; Takakura & Rånby, 1969, 1970).

Experimental

All ESR spectra were taken using a standard Varian V-4502-12 X-band spectrometer equipped with a V-4531 multipurpose cavity bearing a 60-% transmission grid on the front face. The solution sample cells were made from Suprasil to the style of the Varian V-4548 aqueous solution cell but with the internal path length between the two plane faces and the internal breadth being 1.0 and 9.0 mm, respectively. A d.c.-operated, convection air-cooled 1-kW Mazda type ME/D high pressure mercury discharge lamp was employed as the light source, its output being focused on the solution sample cell by way of a two-lens system which included filters to restrict the incident radiation to the region 290–410 nm. The flow-rate range was 0.10– ca. 3.3 ml s^{-1}. Providing the flow rate was at least 0.10 ml s^{-1}, the rise in temperature of the stream in passing through the irradiated zone of the solution sample cell was less than 0.2°C. The experimental arrangements were similar to those of Livingston & Zeldes (1966) and are given elsewhere in more detail (Smith & Stevens, 1972).

Fresh commercial samples of RN:NR, thiophene-free toluene, and the four alkyl methacrylates were used without purification, tests having shown that the RN:NR was at least 99 % pure and that any impurities in the liquids produced no observable effects. Measurements were taken for 0.2 M solutions of RN:NR in toluene, each of the undiluted methacrylic esters, and equivolume mixtures of toluene and each ester. Also, in the case of MMA, runs were carried out with 0.4 M RN:NR solutions in MMA-toluene mixtures covering essentially the entire composition range, the solution sample cells being of internal path length 1.0–2.0 mm. For these particular runs, the light filter arrangement was modified slightly so as to raise the ESR signal level ca. 15%, the lower limit of the flow-rate range consequently having to be raised to 0.12 ml s^{-1} so as to keep the temperature rise of the stream in passing through the irradiated zone at less than 0.2 °C. Furthermore, for each neat ester, test runs were made with the RN:NR omitted and using a solution sample cell of internal pathlength 2.0 mm, no ESR signal being observed. These reaction solutions were first freed of oxygen and then photolyzed at 25 ± 1°C and 39 ± 1°C over the full flow-rate range. For all runs, save those where the light-filter system was modified, the incident light intensity was kept nominally constant.

Results

Photolysis of RN:NR Dissolved in Toluene

This gave rise to the well known spectrum of R· (e.g., Pearson, Smith & Smith, 1964; Brumby, 1970; Bullock, Burnett & Kerr, 1971b) as illustrated in Fig.1. The structure of the spectrum and the coupling constants appeared independent of flow rate and temperature. The signal intensity was marginally dependent on flow rate at constant temperature and on temperature at constant flow rate, e.g., at 25°C, the signal intensity at ca. 3.3 ml s^{-1} was about 5% less than that at 0.10 ml s^{-1}.

Photolysis of RN:NR Dissolved in MMA Undiluted with Toluene

The observed ESR spectra were complex and were found to be essentially independent of flow rate at constant temperature but somewhat dependent on temperature at constant flow rate.

Fig 2B shows a typical spectrum observed at 25°C. Some of the lines are clearly due to R· but the contribution from R· to the whole spectrum is small, viz. about two tenths mole-fraction estimated by comparison of peak heights (Smith et al., 1965). The remaining lines are centered on a *g* value slightly higher than that of R· and may be interpreted as being due to the presence of one other paramagnetic species. The high-field lines are noticeably sharper than the low-field lines, an effect which is well known for certain radicals, e.g., addition radicals formed from substituted ethylenes in aqueous solution (Fischer, 1964a) also, superimposed upon this trend, certain groups of lines are severely broadened even at high field values. This second effect is discussed below.

The radical products expected (Fischer, 1968; Takakura & Rånby, 1969, 1970) from this reaction system are of the type (1) with Y equal to –CH$_3$. The results of previous investigations (Fischer, 1968) of radical addition to this and like methacrylic esters suggest that for such radicals one would expect

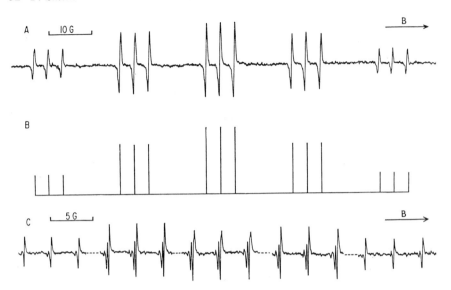

Fig. 1. *A*, First-derivative spectrum for the photolysis of RN : NR in toluene at 25°C; [RN : NR], 0.2 M; flow rate 0.10 ml s⁻¹. Only the central 15 lines are shown. *B*, Stick plot for the central 15 lines of the septet (1 : 6 : 15 : 20 : 15 : 6 : 1) of triplets (1 : 1 : 1) from the R· radical; *C*, The absorption in (*A*) but with the spectrometric recording settings so as to resolve the second-order spectrum.

strong hyperfine interactions from the protons in the β–CH$_3$ and –CH$_2$ groups, and a weak, long-range interaction with the three δ–CH$_3$-group protons (Smith et al., 1965). The observed spectrum agrees with expectation, if the two β–CH$_2$ protons are assumed to be inequivalent, and may be satisfactorily interpreted as a quartet (1:3:3:1) of doublets (1:1) of doublets (1:1) of weak quartets (1:3:3:1). On the basis of this interpretation, the (1:1)-doublets from each of the β–CH$_2$ protons form a quartet (1:1:1:1) with the central pair of lines of each quartet accounting for the broadening phenomenon described above. Few radicals similar to (1) have been characterized in the liquid phase by ESR spectroscopy (Fischer, 1968; Takakura & Rånby, 1969, 1970) and, of these, the most comparable is the analogous polymeric addition radical of methacrylic acid, MAA, formed in aqueous solution and first observed by Fischer (1964a). The non-equivalence of the two β–CH$_2$ protons and broadening of the inner two lines of each β–CH$_2$-group quartet (1:1:1:1) observed for radical (1) is similar to Fischer's (1964) findings for the polymeric addition radical of MAA. This point will be discussed more fully below.

A typical spectrum taken at 40°C is given in Fig. 2*C*. The spectrum from the addition-radical component is essentially the same in every respect as that observed at 25°C save that the broadening of the inner two lines of the (1:1:1:1) quartets is reduced. The resulting improvement of the resolution

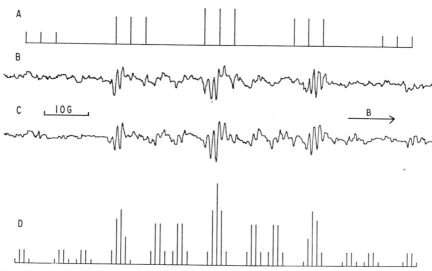

Fig. 2. *A*, Stick plot for the central 15 lines of the spectrum from the R· radical; *B*, First-derivative spectrum for the photolysis of RN:NR in neat MMA at 25°C; [RN:NR], 0.2 M; flow rate 0.10 ml s⁻¹. *C*, Recorded with the use of the same reaction conditions and spectrometer settings as for (*B*) save that the temperature was 40°C; *D*, Stick plot for the quartet (1:3:3:1) of doublets (1:1) of quartets (1:3:3:1) from the terminal structure –CH₂–Ċ(COOCH₃)CH₃.

of these groups of inner lines enabled more reliable measurement of the coupling constants to be made.

Photolysis of RN:NR Dissolved in MMA-Toluene Mixtures

The ESR spectra observed were similar to those reported above for neat MMA. As was observed with the use of neat MMA, the spectra did not change appreciably with flow rate. When the concentration of MMA was less than ca. 1.9 M, the only radical observed was R· but, as the concentration of MMA was gradually increased by roughly 1 M steps, this spectrum was gradually replaced by that assigned to radicals of type (1). The most detailed measurements were carried out at a flow rate of 0.50 ml s⁻¹. Within experimental accuracy, the structure and line positions of the secondary spectrum were independent of the concentration of the MMA and the flow rate but there was present the temperature dependent line width phenomenon noted above. However, the possible significance of these observations is limited by the low intensity of the contribution from the addition-radical component to the total spectrum when the MMA concentration was low.

Photolysis of RN:NR Dissolved in EMA, IBMA, and NBMA and their Mixtures with Toluene

The ESR spectra observed for each monomer were analogous to those found for MMA, as illustrated in Fig. 3 which refers to NBMA, and were interpreted in a like fashion.

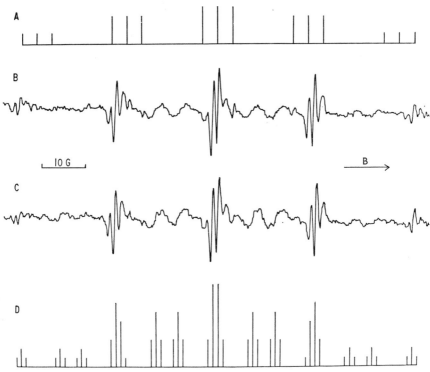

Fig. 3. *A*, Stick plot for the central 15 lines of the spectrum from the R· radical; *B*, First-derivative spectrum for the photolysis of RN : NR in neat NBMA at 25°C [RN : NR], 0.2 M; flow rate 0.27 ml s⁻¹; *C*, Recorded with the use of the same reaction conditions and spectrometer settings as for *B*, save that the temperature was 39°C; *D*, Stick plot for the quartet (1 : 3 : 3 : 1) of doublets (1 : 1) of triplets (1 : 2 : 1) from the terminal structure $-CH_3-\overset{\cdot}{C}$ $(COOCH_2CH_2CH_2CH_3)CH_3$.

The coupling constants and *g* values for all the addition radicals observed are given in Table 1. It will be seen that the comparable data for these radicals are similar. This similarity of the spectroscopic data and the general similarity of all of the results found with these four alkyl methacrylates noted

Table 1. *Hyperfine Coupling Constants and g values[a] at 39 ± 1°C for Addition Radicals of Terminal Structure* $-CH_2-\overset{\cdot}{C}(COOY)CH_3$

Alkyl group, $-Y$	Proton Coupling Constants, G^b		
	$\beta-CH_2$	$\beta-CH_3$	$\delta-COOY$
$-CH_3$	14.18, 9.27	22.19	1.13
$-CH_2CH_3$	14.04, 9.33	22.18	1.19
$-CH_2CH(CH_3)_2$	14.32, 9.32	22.14	1.22
$-CH_2CH_2CH_2CH_3$	14.18, 9.22	22.23	1.19

[a] *g* values were all equal to 2.0034 ± 0.0002.
[b] Maximum uncertainty, ca. 0.05 G.

above is reasonable. This follows because these esters differ to only a moderate extent with respect to the alkoxyl group and such differences in the alkoxyl group would not be expected to greatly affect either the rate constants of the chemical and physical processes taking place in these systems or the ESR spectroscopic constants of the radicals formed. This view is supported, for example, by the available data on the relevant polymerization rate constants (e.g., Ulbricht, 1966) of these alkyl methacrylates and on the coupling constants of related radicals (Fischer, 1964; Takakura & Rånby, 1969, 1970; Smith et al., 1965).

Discussion

The ESR spectra of the addition radicals observed are consistent with the expected structure (1). The magnetic inequivalence of the two β–CH_2 protons is typical for radicals of type

$$X_1CH_2\text{–}\dot{C}(X_2)X_3, \tag{2}$$

where groups X_1, X_2, and X_3 are of such size and shape that there is considerable restriction to rotational motion about the X_1CH_2–$\dot{C}(X_2)X_3$ bond or there is a fixed conformation about that bond (Fischer, 1964, 1968). The observed sharpening of the inner lines of the β–CH_2 quartets with rise in temperature, if interpreted strictly on the basis of the theoretical treatment of Fischer (1964, 1968) for radicals such as (2), seems to support the former possibility. A similar sharpening effect has been noticed already in preliminary studies of the analog of (1) from MAA in aqueous solution (Fischer 1964, 1968). The possible origin of this line-sharpening effect brought about by raising the temperature has been discussed elsewhere (Smith & Stevens, 1972).

The ESR spectroscopic data available in the literature (Fischer, 1968; Takakura & Rånby, 1969, 1970) for radicals of structure analogous to that of (1) do not by themselves enable us to be sure about the average molecular weight of the addition radicals we have observed, for an addition radical such as (1) might have an ESR spectrum which is relatively insensitive to the size of n even when n is small. This follows because, for example, the R-group is sterically and electronically similar to the chain repeating unit. We estimate the steady state radical concentration in our work to be ca. 1×10^{-6} M. This estimate and the kinetic data which apply in our reaction system for the case of long-chain addition radicals of MMA (Shulz, Henrici-Olivé & Olivé, 1960) allow us to calculate the kinetic chain length at 25°C to be ca. 1×10^2.

This calculation of the size of the kinetic chain length does not take into account the possible variation of the chain propagation and termination rate constants with chain length for short-chain radicals and the likely occurrence of primary radical termination. There is little data in the literature to enable us to allow for these factors. However, what data there is suggests that the kinetic chain length at 25°C might be ca. 1×10^1 to ca. 1×10^2 (Smith & Stevens, 1972). The available kinetic data for the other monomers (Ulbricht, 1966) are similar to, although less complete than, the comparable data for MMA, suggesting that a similar conclusion holds for all the monomers studied.

Support for this conclusion is provided by the results of current experimental work on the lines described in this report, but where the azo compounds used were other than RN:NR (Smith, Gilman & DeLorenzo, 1972). This work has shown the spectroscopic data for the addition radicals observed to be independent of the azo compound used.

This work was supported N.S.F. Grants GP 7534 and GP 17579.
The help of Dr. R. D. Stevens, Mr. L. B. Gilman, and Dr. R. A. DeLorenzo is gratefully acknowledged.

References

Brumby, S., Z. Naturforsch., *25a*, 12 (1970).
Bullock, A. T., Burnett, G. M. & Kerr, C. M. L., Eur. Polym. J., *7*, 791 (1971*a*).
Bullock, A. T., Burnett, G. M. & Kerr, C. M. L., Eur. Polym. J., *7*, 1011 (1971*b*).
Czapski, G., J. Phys. Chem., *75*, 2957 (1971).
Doi, Y. & Rånby, B., J. Polymer Sci., Part C, *31*, 231 (1970).
Fischer, H., Z. Naturforsch., *19a*, 866 (1964*a*).
Fischer, H., J. Polymer Sci., Part B, *2*, 529 (1964*b*).
Fischer, H., Adv. Polymer Sci., *5*, 463 (1968).
Flockhart, B. D., Ivin, K. J., Pink, R. C. & Sharma, B. D., J. Chem. Soc., D, 339 (1971).
Hefter, H. & Fischer, H., Ber. Bunsenges. Phys. Chem., *73*, 633 (1969).
Hefter, H. J., Hecht, T. A. & Hammond, G. S., J. Amer. Chem. Soc., *94*, 2793 (1972).
Kawamura, T., Ushio, M., Fujimoto, T. & Yonezawa, T., J. Amer. Chem. Soc., *93*, 908 (1971).
Kochi, J. K. & Krusic, P. J., Chem. Soc. Spec. Publ., *24*, 147 (1970).
Livingston, R. & Zeldes, H., J. Chem. Phys., *44*, 1245 (1966).
Neta, P., Hoffman, M. Z. & Simic, M., J. Phys. Chem., *76*, 847 (1972).
Norman, R. O. C. & Storey, P. R., J. Chem. Soc., B, 1009 (1971).
Oster, G. & Yang, N.-L., Chem. Rev., *68*, 125 (1968).
Pearson, J. T., Smith, P. & Smith, T. C., Can. J. Chem., *42*, 2022 (1964).
Shulz, G. V., Henrici-Olivé, G. & Olivé, S., Z. Phys. Chem. (Frankfurt), *27*, 1 (1960).
Smith, P., Gilman, L. B. & DeLorenzo, R. A., unpublished.
Smith, P., Pearson, J. T., Wood, P. B. & Smith, T. C., J. Chem. Phys., *43*, 1535 (1970).
Smith, P. & Stevens, R. D., J. Phys. Chem., in press.
Smith, P. & Wood, P. B., Can. J. Chem., *45*, 649 (1967).
Takakura, K. & Rånby, B., J. Polymer Sci., Part C, *22*, 939 (1969a).
Takakura, K. & Rånby, B., Advan. Chem. Ser., *91*, 125 (1969b).
Takakura, K. & Rånby, B., J. Polymer Sci., Part A-1, *8*, 77 (1970).

Ulbricht, J., Propagation and Termination Constants in Free Radical Polymerization, Polymer Handbook (ed. J. Brandrup & E. H. Immergut) p. II-57. Interscience, New York (1966).

Discussion

Chapiro

On Fig. 2 *B*, which is for 25°C, the ESR spectrum still contains primary radicals derived from 2,2′-azo-*bis*-isobutyronitrile; it is therefore likely that the average chain length of the addition radicals present is very short ($n < 5$). However, at 40°C (Fig. 2 *C*) primary radicals partly disappear and the lines attributed to long-chain radicals become sharper. This result can be attributed to faster addition to double bonds and therefore to an increase in the average chain length.

P. Smith

Your perceptive observation with respect to Fig. 2 is in agreement with our general experience, e.g., see Fig. 3. We have noticed throughout our work that the value of the ratio $[R·]/[C·]$, where $[C·]$ denotes the sum of the molar concentrations of all the addition radicals present, falls as the temperature is raised while the other significant reaction variables are kept constant. Quantitative study of this effect is hampered by the low intensity of the lines of the $R·$ radicals which may be used. For a given monomer and temperature, determinations of this ratio are found to be reproducible to $\pm 10\%$ and, not surprisingly, the values obtained are about the same size for all four monomers. In the case of MMA, $[R·]/[C·]$ is 0.25 at 25°C and 0.16 at 39°C.

The usual simple mechanism for radical polymerization applied to our reaction systems indicates that the kinetic chain length, v, is given by

$$v = k_p[M]/(2Rk_t)^{\frac{1}{2}} \tag{1}$$

where k_p, k_t, $[M]$, and R are, respectively, the propagation rate constant, the termination rate constant, the monomer molarity, and the initiation rate. From eq. (1) it is reasonable to conclude that the temperature dependence of v is largely determined by that of $k_p/(k_t)^{\frac{1}{2}}$. In the case of MMA, the Arrhenius energy of activation of $k_p/(k_t)^{\frac{1}{2}}$ has been fairly well established by previous workers to be 3–5 kcal mol^{-1}, from which we deduce that $v_{39°}/v_{25°}$ is about 1.3–1.5 if the temperature is varied while keeping the other significant rate variables constant. There are uncertainties in exactly applying eq. (1) to our reaction systems as I will point out later. Nevertheless, it is interesting to note that this numerical result for $v_{39°}/v_{25°}$ tends to support your suggestion that for a given reaction system the average chain length increases on raising the temperature.

Eq. (1) assumes that k_p and k_t are independent of chain length and that primary radical termination is negligible. Of course, in our reaction systems, primary radical termination is not likely to be negligible in general. We can allow for this probability by assuming the R· radical to be of reactivity in addition and termination steps roughly similar to that of the short-chain addition radicals, e.g., that initiation step rate constant k_i is of the same order of magnitude as k_p for short-chain radicals. This assumption is undoubtedly crude. However, it tends to be supported by the kinetic and thermodynamic evidence available for long-chain polymethyl methacrylate and polymethacrylonitrile radicals, e.g., reactivity-pattern and ceiling-temperature data (Smith, P. & Stevens, R. D., J. Phys. Chem., in press). Application of this assumption indicates that eq. (1) would still hold although the meaning of ν would necessarily change slightly. However, we are still left with the problem that both k_p and k_t may vary significantly with chain-length over the chain-length range of possible concern here. In our forthcoming paper (Smith, P. & Stevens, R. D., J. Phys. Chem., in press) we estimate the extent of these variations and their consequences on the size of ν, basing our estimates on the relevant experimental data in the literature. Unfortunately the quantity of such experimental data available is small. It is from such considerations that we estimate ν at 25°C in the case of neat MMA to be ca. 1×10^1 to ca. 1×10^2.

As pointed out in my paper, for each monomer studied, the inner pair of lines of the (1:1:1:1) quartets of the addition-radical component of the ESR spectrum tend to sharpen as the temperature is raised while keeping the other significant reaction variables and the instrumental recording conditions constant. The origin of this effect is discussed in our forthcoming paper (Smith, P. & Stevens, R. D., J. Phys. Chem., in press). One of the factors possibly contributing to this effect considered there is the increase in ν with temperature you have suggested. We are hoping to examine this sharpening phenomenon with the use of azo-compounds other than RN:NR so as to more readily extend our investigations to a temperature range wider than that employed in our present work. By so doing, it may be possible to obtain more definitive data to help our understanding of the source of this interesting effect. Another consequence of raising the temperature is a small improvement in the resolution of the three most prominent bunches of lines from the addition-radical component (e.g., see Figs. 2 and 3). It seems that this may arise because, as the ratio [R·]/[C·] falls, ([C·]+[R·]) remains roughly constant so that the overlap between the spectra from R· and C· which occurs where these three bunches of lines are situated becomes less important.

Charlesby

I would like to ask what evidence there is that the rate constant of addition is independent of chain length. Is the first addition to the primary radical as probable as addition to a longer growing chain radical?

P. Smith

Let me try to start by answering the first part of the question. In reaction media similar to those of our work, i.e. at about room temperature in liquids with low viscosity and high solvent power for the polymer, there is general expectation on theoretical grounds that the rate constant k_p for the addition step in a simple radical homopolymerization such as those described in my paper will be essentially independent of chain length, providing the chain length is greater than ca. 5. However, when the chain-length is less than ca. 5, two important effects are expected to make this generalization break down. The first possibility is when the initiating radical is unusually bulky or bears groups which can exert powerful inductive or electromeric effects. Such effects will influence the reactivity of the propagating radical to a rapidly diminishing extent with increase in chain-length. The second possibility stems from the well-established cyclic rule of 6. The operation of this rule will cause an addition radical able to form an intramolecular 6-membered ring to have an unusual value of k_p.

The validity of these ideas appears to have been generally substantiated by the available experimental results obtained by classical methods (e.g., Bamford, C. H., Barb, W. G., Jenkins, A. D. & Onyon, P. F, in The Kinetics of Vinyl Polymerization by Radical Mechanisms, Chapt. 3. Butterworth, London 1958). Some of the most recent work in this area has been concerned with polymerizations in the presence of bromotrichloromethane which can act both as a photoinitiator and as a powerful chain-transfer agent (Bengough W. I. & Thomson, R. A. M., Trans. Faraday Soc., *61*, 1735 (1965); Barson, C. A., Mather, R. R. & Robb, J. C., Trans. Faraday Soc., *66*, 2585 (1970)). In these studies the short-chain polymer radicals which could be kinetically studied bear the strongly polar initiator fragment $-CCl_3$ which would be expected to be relatively reactive to intramolecular attack *via* the cyclic mechanism already described. Further supporting evidence is provided by the few kinetic investigations by ESR spectroscopy so far carried out (e.g., Fischer, H., Makromol. Chem., *98*, 179 (1966)).

The results of these various studies indicate that k_p for short-chain addition radicals may be one or possibly two powers of ten different from that for long-chain addition radicals. However, the amount of kinetic work done on short-chain addition radicals is very small and of limited scope. This is unfortunate, for it will be apparent that, for a given monomer, the extent of the variation of k_p with chain-length will probably be sensitive to the identity of the initiator fragment. As far as I am aware, reliable experimental results easily related to those of our study are not available.

In answer to the second part of the question, as discussed in our forthcoming paper (Smith, P., & Stevens, R. D., J. Phys. Chem., in press), there are reasonable arguments for regarding the R· radical as the first member of the addition-radical series. These arguments are based, for example, on the similarity of long-chain polymethyl methacrylate and polymethacrylonitrile radicals with respect to their kinetic and thermodynamic properties. In which case, the discussion about the variation of size k_p with chain length which I gave in the earlier part of my answer would include the size of the rate constant k_i for the first addition step. Incidentally, the arguments given in that discussion suggest that k_p in our reaction systems may not vary with chain-length as much as the maximum extent quoted, i.e., because the R· radical is roughly similar to the $-CH_2-C(COOY)CH_3-$ repeated unit in the addition-radical chain with respect to steric and electronic properties.

Chapiro

In answer to the question of Professor Charlesby, I would like to draw attention to work carried out in Melville's laboratory many years ago, where it was shown, using the rotating sector method, that the rate constant for addition to the double bond k_p decreased with chain-length but only for the first members of the series. For chain-lengths higher than ca. 5 or 6, k_p becomes independent of chain-length of the radical. In the present study, the primary radical R· derived from AIBN (RN:NR) is perhaps less reactive with respect to the double bond than is the chain propagating radical.

P. Smith

I am not personally aware of the work of Melville and his coworkers which you quote. Clearly, it does seem to fit into the general picture I have painted in my reply to Professor Charlesby. However, unless the data of the Melville group were for reaction systems closely similar to those of our studies, their results might not be very applicable to a detailed interpretation of our results. Of course, this does not rule out the possibility that the rate constant for the initiation step might be less than, say, that for the propagation step for the radical $RCH_2\dot{C}(COOY)CH_3$. Certainly, it would seem to be fortuitous if these rate constants were essentially equal.

Lindberg

Have you calculated the energy of activation of the propagation processes taking place in any of your reaction systems?

P. Smith

This question and related matters are dealt with in detail in our forthcoming paper (Smith, P. & Stevens, R. D., J. Phys. Chem., in press). The approach there has been to show first that our ESR measurements are probably on reaction mixtures largely in the stationary state. Then we make use of our results for the dependence on temperature under otherwise constant reaction conditions of the stationary-state value of the quantity F equal to $[R\cdot]/([C\cdot]+[R\cdot])$ where $[C\cdot]$ represents the sum of the concentrations of all the addition radicals present, i.e. $\sum_{x=1}^{\infty}[R(M)_x\cdot]$ where M is the monomeric repeating unit and $R(M)_x\cdot$ stands for an addition radical such as (1) in my paper (p. 29). We assume the usual simple mechanism of radical polymerization, viz. the photodissociation step

$$RN:NR + h\nu \rightarrow 2R\cdot + N_2 \tag{1}$$

plus initiation, propagation, and termination steps with rate constants k_i, k_p, and k_t, respectively, k_p and k_t being assumed independent of radical size. Also, we make the crude assumption that the rate constants for the reactions of the $R\cdot$ radical are equal to the corresponding rate constants for the $R(M)_x\cdot$ radicals, i.e. k_i and k_p are approximately equal and the rate constant for the primary radical termination steps

$$R\cdot + R(M)_x\cdot \rightarrow product(s) \tag{2}$$

is approximately equal to k_t. From these assumptions, we show that the stationary-state value of F depends on the kinetic chain length, ν, by way of the equation

$$F = \nu^{-1}/(1 + \nu^{-1}) \tag{3}$$

and we argue that it is reasonable to conclude that the temperature dependence of ν is probably close to that of $k_p/(k_t)^{\frac{1}{2}}$. The application of these ideas to our reaction systems is then illustrated by considering the case of MMA for which there is the most extensive and reliable kinetic data in the literature. These data indicate that the overall Arrhenius energy of activation of $k_p/(k_t)^{\frac{1}{2}}$ is 3–5 kcal mol^{-1} and, as also stated in my present paper, ν at 25°C is between ca. 1×10^1 and ca. 1×10^2. Using this information in eq (3) gives $F_{39°}/F_{25°}$ to be ca. 0.8. Such good agreement is regarded as somewhat encouraging.

Rånby

My comment to Prof. P. Smith's paper is that the DP values derived (~ 100)

appear rather high to me. To obtain sufficiently high radical concentration for the recording of ESR spectra (10^{-6}–10^{-5} M), the initiation rate is so high that a high rate of termination with initiation radicals would also be expected. In a flow system, the DP of the polymer radicals would probably not exceed 5–10.

Free Radical Initiation Reactions of Allyl, Methallyl and 1,2-Substituted Monomers

By Zenzi Izumi and Bengt Rånby

Department of Polymer Technology, The Royal Institute of Technology, Stockholm, Sweden

This report describes a part of the research program on free radical polymerization carried out in these laboratories since 1964 (Takakura & Rånby, 1969).

In general, allyl and methallyl monomers polymerize with low rate and give low degree of polymerization. 1,2-substituted unsaturated compounds have not generally been homopolymerized but copolymerized with other monomers. Not much is known about the mechanism of allyl polymerizations and the reactions of 1,2-substituted unsaturated compounds, which are complex reactions. ESR measurements using the flow technique are particularly useful for initiation reaction studies of allyl, methallyl and 1,2-substituted monomers. Previously, however, only allyl alcohol (Smith & Wood, 1967), allyl amine (Corvaja, Fischer & Giacometti, 1965), trimethylolpropane monoallyl ether (Doi & Rånby, 1970), maleic acid (Dixon & Norman, 1964) and fumaric acid (Fischer, 1964) have been studied using this method. We report here reactions of three types of free radical initiators ($HO\cdot$, $H_2N\cdot$ and $H_3C\cdot$) with eight types of allyl and methallyl monomers and four types of crotonic compounds (Table 1).

The reactivity of monomers to various radicals, polymerizability of monomers, the structure and steric conformation of the transient monomer radicals formed, and the information on coupling constants of substituents have been derived from the ESR measurements.

The experiments were carried out with a flow apparatus attached to the ESR instrument, as previously described (Yoshida & Rånby, 1965; Takakura & Rånby, 1970). Two aqueous solutions were used in each case, i.e., oxidizing solution of H_2O_2, $NH_2OH \cdot HCl$ or $(CH_3)_3COOH$, respectively, containing a monomer (usually about 1.1×10^{-1} mol l^{-1}) and one reducing solution of $TiCl_3$. Both solutions contained either sulfuric acid (to pH 1.4) or ethylenediaminetetraacetic acid (EDTA) and a phosphate buffer (to pH 7).

Table 1. *List of Allyl, Methallyl and Crotonic Compounds studied*

Substrate	Formulae
Allyl alcohol (AA)	$CH_2 = CH \ CH_2OH$
Allyl amine (AAm)	$CH_2 = CH \ CH_2NH_2$
Allyl acetate (AAc)	$CH_2 = CH \ CH_2OCOCH_3$
Sodium allyl sulfonate (SMAS)	$CH_2 = CH \ CH_2SO_3Na$
Methallyl alcohol (MAA)	$CH_2 = C(CH_3)CH_2OH$
Methallyl amine (MAAm)	$CH_2 = C(CH_3)CH_2NH_2$
Methallyl acetate (MAAc)	$CH_2 = C(CH_3)CH_2OCOCH_3$
Sodium methallyl sulfonate (SMAS)	$CH_2 = C(CH_3)CH_2SO_3Na$
Crotonyl alcohol (CAL)	$CH_3CH = CH \ CH_2OH$
Crotonic acid (CAC)	$CH_3CH = CH \ COOH$
Crotonitrile (CNI)	$CH_3CH = CHCN$
Crotonaldehyde (CAD)	$CH_3CH = CHCHO$

Results and Discussion

Among many experiments, the spectra of allyl acetate (AAc) with H_2O_2–$TiCl_3$, sodium methallyl sulfonate (SMAS) with three types of free radical initiators, crotonic acid (CAC) with NH_2OH–$TiCl_3$ and crotonitrile (CNI) with H_2O_2–$TiCl_3$ are shown here (Figs. 1–6) to demonstrate the experimental results obtained.

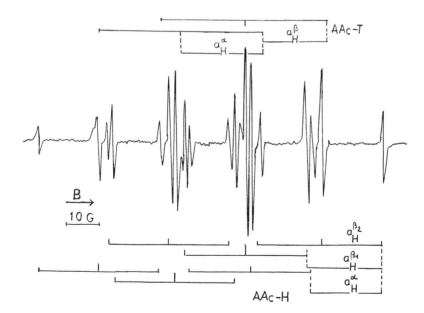

Fig. 1. ESR spectrum from allyl acetate (AAc) in the system H_2O_2–$TiCl_3$: H, monomer head radical, T, monomer tail radical [AAc] $= 5.5 \times 10^{-2}$ mol l^{-1}, [H_2O_2] $= 1.1 \times 10^{-2}$ mol l^{-1}, [$TiCl_3$] $= 8 \times 10^{-3}$ mol l^{-1}, [H_2SO_4] $= 2.2 \times 10^{-2}$ mol l^{-1}.

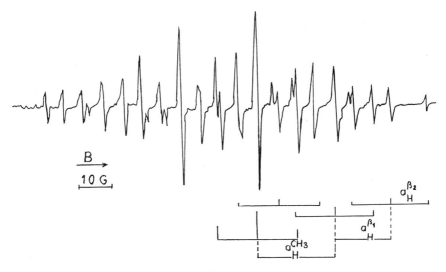

Fig. 2. ESR spectrum from sodium methallyl sulfonate (SMAS) in the system H_2O_2–$TiCl_3$: $[SMAS] = 5.5 \times 10^{-2}$ mol l^{-1}, $[H_2O_2] = 1.1 \times 10^{-2}$ mol l^{-1}, $[TiCl_3] = 8 \times 10^{-3}$ mol l^{-1}, $[H_2SO_4] = 2.2 \times 10^{-2}$ mol l^{-1}.

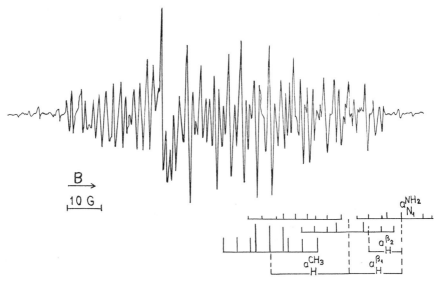

Fig. 3. ESR spectrum from SMAS in the system $NH_2OH \cdot HCl$–$TiCl_3$: $[SMAS] = 1.1 \times 10^{-1}$ mol l^{-1}, $[NH_2OH \cdot HCl] = 0,25$ mol l^{-1} $[TiCl_3] = 4 \times 10^{-3}$ mol l^{-1}, $[H_2SO_4] = 2.2 \times 10^{-2}$ mol l^{-1}.

Reactivity of Monomers to Various Initiating Radicals

Allyl and Methallyl Monomers

When $H_2N \cdot$ was used as initiating radical, only monomer head radicals (e.g. H_2N–CH_2–$\dot{C}H$–CH_2–R) could be obtained for the allyl and methallyl monomers studied.

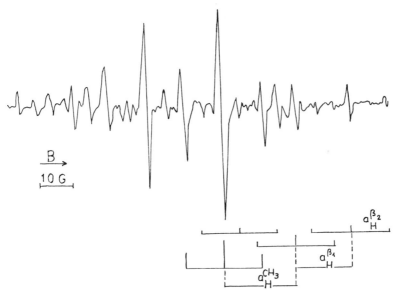

Fig. 4. ESR spectrum from SMAS in the system t-butylhydroperoxide (BHP)–TiCl$_3$: [SMAS] = 5.5×10^{-2} mol l^{-1}, [t-BHP] = 0.50 mol l^{-1}, [TiCl$_3$] = 8×10^{-3} mol l^{-1}, [H$_2$SO$_4$] = 2.2×10^{-2} mol l^{-1}.

It is reasonable to assume that the protonated amino radicals are strongly electrophilic and tend to react with the monomer at the furthest position from substituents with -1 effect (acetate groups are an exception). A contributing reason for getting only head radicals could be that the protonated amino radicals are too bulky for an attack on α or allyl carbons due to steric hindrance which may explain the results with allyl acetate.

In the cases where HO· is used as an initiating radical, the results are rather complex. In addition to head radicals (the main product), appreciable amounts of monomer tail radicals (e.g. ·CH$_2$–CHOH–CH$_2$–R) were formed with four allyl monomers. Formation of tail radicals in our experiments could be due to the electrostatic properties, and the small size of HO· radicals, the expected low resonance stabilization of head radicals, and the electrostatic and steric properties of monomers. It is remarkable that only α-methylene radicals could be detected in the case of methallyl acetate (MAAc). The H$_3$C radicals have a low reactivity (Michewich & Turkewich, 1968) and add only to positively polarized reactive double bond, i.e. in AA and SMAS.

Crotonic Compounds

Generally, two types of radical (A, B) could be obtained for crotonic compounds (generally written ^{1}CH$_3$–^{2}CH = ^{3}CH–^{4}CR) by reacting with various radicals, except for crotonaldehyde. Radical spectrum A is assigned to the radicals

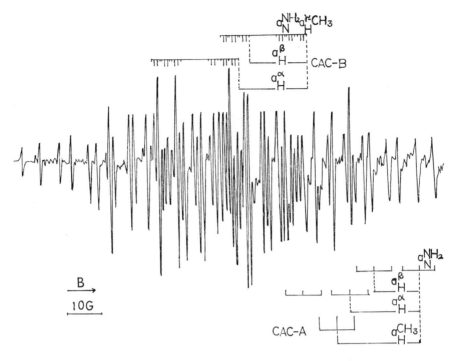

Fig. 5. ESR spectrum from crotonic acid (CAC) in the system NH$_2$OH·HCl·TiCl$_3$:

$$\text{A} \quad H_2N-\underset{\underset{COOH}{|}}{\overset{\overset{H}{|}}{C}} - \underset{\underset{CH_3}{|}}{\overset{\overset{H}{|}}{C}} \cdot \qquad \text{B} \quad H_2N-\underset{\underset{CH_3}{|}}{\overset{\overset{H}{|}}{C}} - \underset{\underset{COOH}{|}}{\overset{\overset{H}{|}}{C}} \cdot$$

[CAC] = 5.5 × 10^{-2} mol l^{-1}, [NH$_2$OH·HCl] = 0.25 mol l^{-1}. [TiCl$_3$] = 4 × 10^{-3} mol l^{-1}, [H$_2$SO$_4$] = 2.2 × 10^{-2} mol l^{-1}.

(CH$_3$–ĊH–CH(R′)–CR) produced by the addition of initiator radicals R′ to carbon atom 3 substituted with a functional group. Radical spectrum B is assigned to the radicals (CH$_3$–CH(R′)–ĊH–CR) produced by the addition of initiator radicals to the carbon atom 2 substituted with a methyl group.

The concentration ratio of the two types of radicals depends on the substituents of the crotonic compounds and the type of attacking radicals. As for the substituent CR of crotonic compounds, the ratio of A type radical decreases in the order CH$_2$OH > COOH > CN > CHO, which may be related to the resonance effect of the substituents. As for the attacking radicals, the ratio of A type radical decreases generally in the order H$_3$C > HO > H$_2$N, which shows that a nucleophilic radical attacks the carbon atom with an electron-withdrawing substituent more than that with an electron-donating substituent as CH$_3$.

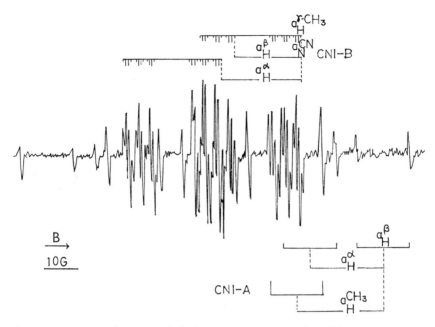

Fig. 6. ESR spectrum from crotonitrile (CNI) in the system H_2O_2–$TiCl_3$:

$$
\text{A} \quad \begin{array}{c} H \quad H \\ HO—C — C \\ | \quad | \\ COOH \ CH_3 \end{array}
\qquad
\text{B} \quad \begin{array}{c} H \quad H \\ HO—C — C \\ | \quad | \\ CH_3 \ COOH \end{array}
$$

$[CNI] = 5.5 \times 10^{-2}$ mol l^{-1}, $[H_2O_2] = 1.1 \times 10^{-2}$ mol l^{-1}. $[TiCl_3] = 8 \times 10^{-3}$ mol l^{-1}, $[H_2SO_4] = 2.2 \times 10^{-2}$ mol l^{-1}.

Polymerizability of Monomers

Allyl and Methallyl Monomers

There are no indications of polymer radicals in the recorded spectra, although rather high monomer concentrations have been used. The low reaction rate and generally low degree of polymerization of allyl polymers have been considered to be due to formation of allylic type radicals. It is, therefore, unexpected that allylic type radicals are found only in the case of AA. More detailed studies seem necessary. This study has indicated that there may exist some reason for the low rate and the low degree of polymerization of allyl compounds in the existence of tail type radical and very low reactivity of both monomer and monomer radical, which is shown in the study of copolymerization (Izumi & Rånby, unpublished results).

Crotonic Compounds

The degree of steric hindrance of monomers is clearly related to the value θ, which can be calculated (Fischer, 1967) from the equation

$$a_H^\beta = \varrho B_H^\beta \left(\frac{3 - 2\cos^2\phi}{4}\right) = \varrho B_H^\beta \overline{\cos^2\theta}$$

where θ is the angle between the projection of the C^β–R bond and the axis of the $2p_z$ orbital. θ values of all crotonic compounds lie between $17°$–$23°$ when R is HO. These monomers can not be homopolymerized but can be copoly-merized with less steric hindered monomer such as acrylonitrile. θ values of all monomers so far studied that can homopolymerize are higher than $26°$, with the exception of vinyl esters, where special effects have been indicated (Takakura & Rånby, 1970). θ values may be used as a measure of polymeriza-bility when it is related to steric hindrance.

Steric Conformation of Observed Radicals

The coupling constants of β protons vary considerably with the nature of the substituents. The β_1 coupling constants decrease in the order $HO > H_3C > H_2N$ and the β_2 coupling constants (allyl hydrogen) decrease in the order $CH_2OH > CH_2NH_2 > CH_2OCOCH_3 > CH_2SO_3Na$ for allyl and methallyl compounds, i.e. in the reverse order of bulkiness of the groups. The same tendency is indicated for crotonic compounds. The β coupling constants decrease in the order $HO > H_3C > H_2N$ and $CN > CHO > CH_2OH > COOH$. Some exceptions could be explained as due to complex formation with Ti^{4+} or by an intramolecular hydrogen bonding between initiator fragment and the substituents at α carbon atom.

γ–CH_3 Splittings and Nitrogen Splittings

The coupling constants of γ–CH_3 protons decrease in the order $COOH > CHO > CH_2OH > CN$ and $HO > H_3C > H_2N$. It is evident that the spin density on the γ–CH_3-groups can be decreased by substituents which contribute to the hyperfine splitting.

The value of a_N in the series of $R—\overset{\displaystyle H}{\underset{\displaystyle CH_3}{C}}—\overset{\displaystyle H}{\underset{\displaystyle CN}{C}}\cdot$ is dependent upon the nature of R.

σ_N varies with R in the order $H_3C > HO > H_2N$, i.e. is decreased by electron-withdrawing substituents as shown (Lemoine et al., 1965) for the series $Me_3C–N(O^-)–Ar$, where Ar is a parasubstituted benzene ring.

The studies presented in this paper are part of a research project on radical polymeriz-ation supported by the Swedish Board for Technical Development.

References

Corvaja, C., Fischer, H. & Giacometti, G., Z. Phys. Chem., *45*, 1 (1965).
Dixon, W. T., Norman, R. O. C. & Buley, A. L., J. Chem. Soc., *1964*, 3625.
Doi, Y. & Rånby, B., J. Polymer Sci., C, 231 (1970).

Fischer, H., Z. Naturforsch., *19a*, 866 (1964).

Izumi, Z. & Rånby, B., unpublished results.

Lemoine, H. et al., Bull. Soc. Chim. France, *1965*, 372.

Michewich, D. & Turkewich, J., J. Phys. Chem., *72*, 2703, (1968).

Smith, P. & Wood, P. B., Can. J. Chem., *45*, 649 (1967).

Takakura, K. & Rånby, B., Adv. Chem. Series (ACS), No *91*, 125–44 (1969).

Takakura, K. & Rånby, B., J. Polymer Sci., A-1, *8*, 77 (1970).

Yoshida, H. & Rånby, B., in Macromolecular Chemistry, Prague 1965 (J. Polymer Sci., C, *16*) (ed. O. Wichterle & B. Sedlacek) p. 1333 (part 1). Interscience, New York 1967.

Discussion

D. R. Smith

You report that hydroxyl radicals abstract hydrogen from allyl alcohol but add to the other allyl derivatives. Similar abstraction reactions have been noted in the literature. For example photolysis of H_2O_2 in allyl alcohol results in abstraction. Gaseous hydrogen atoms have been observed to abstract from allyl alcohol at low temperatures and Dr Tench and I obtained a similar result for allyl alcohol absorbed on MgO. I am curious why you do not see abstraction by hydroxyl from the other allyl derivatives.

Rånby

The hydroxyl group and the vinyl group both tend to activate the alpha C–H bond, facilitating abstraction in the case of allyl alcohol.

Chapiro

In addition, there is a significant energy gain in the formation of an H–O bond in water, the product of abstraction.

P. Smith

The other allyl derivatives, reported in this work, favour addition because substituents such as COOH are considered to tend to deactivate the alpha C–H bond. Your results that the allyl monomers all undergo hydroxyl–radical addition at both ends of the double bond and amino-radical addition only at the unsubstituted end are similar in general to the results we (P. Smith, R. A. Kaba, W. J. Maguire & P. B. Wood, unpublished work; P. Smith & P. B. Wood, Can. J. Chem. *45*, 649 (1967)) have observed with the use of the $TiCl_3$–H_2O_2 and –NH_2OH radical-generating systems at 25°C. E.g., Dr Maguire and I have found propene to give the head radical as the major product and the tail radical as the minor product with the use of the $TiCl_3$–H_2O_2 system; while only the head radical was observed with the use of the $TiCl_3$–NH_2OH system. Propene has recently been studied at low temperature by Hammond and coworkers (H. J. Hefter, T. A. Hecht and G. S. Hammon, J. Amer. Chem. Soc., **94**, 2193 (1972)) with the use of hydroxyl radicals photochemically generated from hydrogen peroxide. They also find the head radical as the major product and the tail radical as the minor product.

Izumi

We obtained hydrogen abstracted type radicals also in the case of crotyl alcohol. Therefore, it may be the general tendency for a hydroxyl group to activate the alpha C–H bond.

It requires more study to know the reason why we could not get hydrogen abstracted type radicals from other allyl monomers. This fact is inconsistent with the general understanding that the low reaction rate and generally low degree of polymerization of allyl polymers is due to formation of allylic type radicals.

Ivin

With our radical system (see paper p. 17) we do not observe radicals formed by hydrogen abstraction from allyl alcohol. It may be that hydrogen abstraction occurs in our system but is immediately followed by addition of sulphur dioxide. Almost certainly different radicals will have different reactivity towards allyl alcohol, both with respect to addition and hydrogen abstraction.

ESR Studies of Copolymerization

By Bengt Rånby

Department of Polymer Technology, The Royal Institute of Technology, Stockholm, Sweden

The empirical theory for copolymerization of vinyl monomers with free radical mechanism was developed in the 1940-ies (cf. Alfrey, Bohrer & Mark, 1952; Ham, 1964, for reviews). Until now the interpretation has largely been based on measurements of reaction rates and analysis of molecular weight and monomer composition of the resulting copolymers. As a new approach during the 1960-ies the amounts, structure and reactions of the transient short-lived free radical species have been made available for direct measurements by application of ESR spectroscopy to polymerization using flow systems. As part of a research program on free radical polymerization (Takakura & Rånby, 1969 a), ESR studies of copolymerization have been made (Takakura & Rånby, 1967, 1969 b), using the same experimental arrangements as previously described (Yoshida & Rånby, 1967).

Experimental

The experiments were made in a flow apparatus of the type developed by Dixon & Norman (1963). The free radical spectra were observed during polymerization when the reacting solutions flowed through a flat quartz cell, 0.25 mm thick, inserted in a TE_{102} rectangular resonant cavity of an X-band ESR spectrometer (Jeol Co. Ltd, Model JES-3B). The polymerization was initiated with hydroxyl radicals formed in a redox reaction between aqueous solutions of hydrogen peroxide (0.22–0.30 M) and titanous chloride (0.14–0.16 M). The pH of both solutions was adjusted to 1.2–1.4 with sulfuric acid or to 7.0 with EDTA and phosphate buffer. The two aqueous solutions, containing the same concentration of monomer or monomer mixture, were mixed immediately before they entered the flat cell in the cavity. The solutions were passed through the cell with gravity feed, giving a flow rate of about 4.0 ml s^{-1}. The dead volume between the mixing junction and the entrance of the flat part of the cell was about 0.06 ml which gave a time lag of about 0.015 s^{-1}. The ESR measurements were carried out at room temperature, e.g. $22 \pm 2°C$. The magnetic field was calibrated with the proton NMR signals. The relative concentrations of radicals were measured from the peak

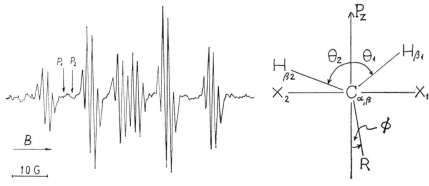

Fig. 1   Fig. 2

Fig. 1. ESR spectrum of vinyl acetate monomer radicals, initiated by addition of HO· radicals. [VAc] = 0.055 mol l⁻¹, [H₂O₄] = 0.11 mol l⁻¹, [TiCl₃] = 0.007 mol l⁻¹, [H₂SO₄] = 0.022 mol l⁻¹. P_1 and P_2 indicate positions for HO· complex radical spectra.

Fig. 2. Steric conformation of the free radical of the type $R-CH_2-CX_1X_2$ with the $C_\alpha-C_\beta$ bond perpendicular to the paper plane.

heights and widths of the spectral components. The total concentration of free radicals observed in these experiments was estimated to be in the range of 10^{-6} to 10^{-5} mol l⁻¹.

Binary Monomer Systems Studied

Two different types of monomer systems have been chosen for these studies. (a) Vinyl acetate (VAc) as a reference monomer (M_1) of low reactivity, combined with a series of comonomers (M_2): acrylonitrile (AN), acrylic acid (AA), maleic acid (MA), and fumaric acid (FA).
(b) Acrylonitrile (AN) as a reference monomer (M_2) of rather high reactivity combined with another series of comonomers (M_1): vinyl acetate (VAc), allyl alcohol (AA), allyl acetate (AAc), sodium vinyl sulfonate (SVS), sodium methallyl sulfonate (SMAS),

(Note that AA and MA refer to different monomers in the (a) and (b) series.)

VAc has a negatively polarized double bond and reacts, therefore, easily with HO· radicals as initiator. The free radical formed (HO–VAc·) reacts preferentially with the comonomers mentioned in (a), and only to a small extent with its own monomer. Three radicals dominate, when a mixture of VAc with a comonomer is initiated with HO· radicals: VAc·, M_2· and copolymer radicals VAc–M_2·. The experiments measure the reactivity of the M_2 monomers to VAc· radicals.

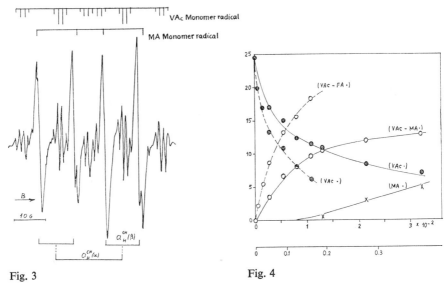

Fig. 3

Fig. 4

Fig. 3. ESR spectrum of the polymerizing system of VAc and MA in acid aqueous solution. The copolymer radical VAc–MA· is predominant. [VAc] = 0.055 mol l⁻¹, [MA] = 0.021 mol l⁻¹.

Fig. 4. *Abscissa:* (outer) Mole fraction of MA (or FA) in VAc–MA (or VAc–FA) system; (inner) [maleic acid] or [fumaric acid] mol l⁻¹; *ordinate:* rad. conc. (arb. unit).

Concentration of different radicals measured from ESR spectra during copolymerization of VAc with MA and VAc with FA, respectively. [VAc] = 0.055 mol l⁻¹, [MA] and [FA] varies. Monomer radicals (VAc·, MA·) and copolymer radicals (VAc–MA·, VAc–FA·) are indicated.

The AN monomer mixtures react differently. Both monomer radicals (AN· and M_1) form easily but the reactivity of the AN· radicals is low due to resonance stabilization. The dominant copolymer radicals are, therefore, M_1–AN·. The experiments measure in this case the reactivity of AN to the M_1 monomer radicals. An added complication with the allyl comonomers is their ability to form two different alkyl radicals (by addition of HO· to the tail and the head of the monomer, respectively) and one allyl radical (by abstraction of a hydrogen).

Copolymerization of Vinyl Acetate (VAc)

The VAc monomer radical has a well resolved ESR spectrum (Fig. 1) described as a doublet of triplets of quartets and interpreted as due to hyper-fine coupling to H_α, H_β and H_{CH_3} respectively.[1]

$$HO-CH_3'-\dot{C}H-OCOCH_2'' \qquad (1)$$
$$\beta \alpha$$

The H_β-coupling constant (a) has a magnitude related to the conformation of the radical as determined by the restricted rotation in the $C_\alpha - C_\beta$ bond.

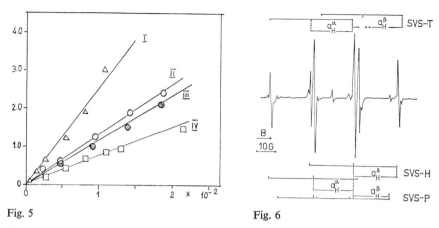

Fig. 5   Fig. 6

Fig. 5. *Abscissa:* [M_2]/mol l^{-1}; *ordinate:* [VAc–M_2·]/[Vac·].
  Plots of concentration ratios [VAc–M_2·]/[VAc·] versus concentration of comonomer M_2 for the systems: I, VAc–FA; II, VAc–AN; III, VAc–AA; IV, VAc–MA with [VAc] = 0.055 mol l^{-1}.

Fig. 6. ESR spectrum of sodium vinyl sulfonate (SVS) initiated with HO· at pH 1.4 and [SVS] = 0.055 mol l^{-1}. Spectra of head (H), tail (T) and polymer radicals (P) are indicated.

For aliphatic radicals, the $a_{H\beta}$ values are given by a relationship

$$a_{H\beta} = B_{H\beta} \varrho_\alpha \cos^2 \theta \quad (B_{H\beta} \text{ is defined below, } \varrho_\alpha \text{ is the spin density on } C_\alpha) \quad (2)$$

derived by Heller & McConnell (1960) where θ is the angle between the axis of the $2p_z$ orbital of the unpaired electron and the direction of the C_β–H bond, projected on a plane perpendicular to the direction of the C_α–C_β bond (Fig. 2). If the two β protons are equivalent, $a_{H\beta}$ can be calculated by the equation of Fischer (1964)

$$a_{H\beta} = B_{H\beta} \varrho_\alpha \tfrac{1}{4}(3 - 2 \cos^2 \phi) = B_{H\beta} \varrho_\alpha \overline{\cos^2 \theta} \quad (3)$$

where ϕ is the angle of free rotation between the projection of the C_α–R bond and the axis of the $2p_z$ orbital. The angle θ represents the average position of the substituent R projected on a plane perpendicular to the C_α–C_β bond as illustrated in Fig. 2, $\overline{\cos^2 \theta}$ is an average for all angles θ attained, and $B_{H\beta}$ (= 58.6 G) is a quantity reported by Fessenden & Schuler (1963) from measurements on simple alkyl radicals.

  With increasing bulkiness and polarity of R (Fig. 2), the angle ϕ decreases. Polymer and copolymer radicals show lower ϕ values, i.e. lower H_β coupling, than the corresponding monomer radicals. This is one of the basic data for the analysis of ESR spectra for polymerizing systems. Vinyl ester monomer radicals have so low $a_{H\beta}$ values that a completely locked conformation

Table 1. *Conversion Rates of* $(M_1^{\cdot})$ *to* $(M_1 - M_2^{\cdot})$ *Radicals from ESR Data*

System M$_1$	M$_2$	Slope of $[M_1 - M_2^{\cdot}]/[M_1^{\cdot}]$ vs. $[M_2]$ Conversion Rate l mol^{-1}	$1/r_1$	(Q, e) Values for M$_2$ Q	e
VAc	Maleic Acid	0.8×10^2	8.3	0.05	1.27[a]
VAc	Acrylic Acid	1.2×10^2	10.0	1.15	0.77
VAc	Acrylonitrile	1.3×10^2	16.4	0.60	1.20
VAc	Fumaric Acid	2.5×10^2	43.5	0.76	1.49[a]

[a] (Q, e) values for the corresponding methyl esters.

$(\phi = 0°)$ is indicated from eq. (3). This is interpreted as an effect of hydrogen bonding of these monomer radicals to a ring structure (Takakura & Rånby, 1970). Vinyl ester polymer radicals have larger $a_{H\beta}$ values which would mean an unlocked structure. This is reasonable, considering the steric conditions for hydrogen bonding.

The initiation of a monomer mixture of VAc and maleic acid (MA) gives an ESR spectrum containing three spectral components: VAc$^{\cdot}$ monomer radicals, and MA$^{\cdot}$ monomer and VAc–MA$^{\cdot}$ copolymer radicals, of which the latter two are largely overlapping (Fig. 3). The $a_{H\beta}$ splitting of the copolymer radicals is smaller (11.1 G) than that of the MA$^{\cdot}$ monomer radicals (12.7 G). Concentration data for a series of measurements of radical concentrations during copolymerization of VAc with maleic acid (MA) and fumaric acid (FA) are given in Fig. 4.

Corresponding experiments for copolymerization of VAc with acrylonitrile (AN) and acrylic acid (AA) were made and the resulting data are summarized in Fig. 5 for the four monomer pairs. Of the four comonomers, MA has the lowest and FA the highest reactivity to VAc$^{\cdot}$ radicals. This is well in line with reciprocal reactivity ratios r_1 $(r_1 = k_{11}/k_{12})$ for the monomer pairs, derived from conventional copolymerization studies and reported in the literature (Table 1).

Copolymerization of Acrylonitrile (AN)

Some of the comonomers used in this series form more than one radical species during the initiation (Izumi & Rånby, 1972). The ESR spectrum of sodium vinyl sulfonate (SVS) indicates one dominant species, interpreted as head radicals HO–CH$_2$–ĊH–SO$_3$ (Na, H). In addition, tail radicals and polymer radicals are formed (Fig. 6). Initiation of a mixture of SVS and AN gives an ESR spectrum showing the two monomer radicals and a copolymer radical interpreted as SVS-AN (Fig. 7). A series of copolymerization experiments with increasing

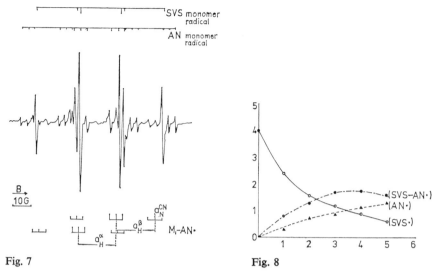

Fig. 7   Fig. 8

Fig. 7. ESR spectrum of a copolymerizing system of SVS and acrylonitrile (AN) at pH 1.4, [SVS] = 0.055 and [AN] = 0.020 mol l⁻¹, showing two monomer and one copolymer radical.

Fig. 8. *Abscissa:* [AN]/mol l⁻¹ × 10²; *ordinate:* rad. conc. (arb. unit).
Radical concentrations measured from ESR spectra during copolymerization of SVS (0.055 mol l⁻¹) with AN at different concentrations.

amounts of AN is summarized in Fig. 8. No indication of AN-SVS· copolymer radicals was found.

Allyl alcohol (AA) forms three different radicals when initiated with HO· radicals:

head radicals $HO-CH_2-\dot{C}H-CH_2OH$ (H)

tail radicals $HO-CH(CH_2OH)-\dot{C}H_2$ (T)

allyl radicals $CH_2{=}\dot{C}H{=}CH-OH$ (A)

of which the head radicals (H) dominate. When a mixture of AA and AN is initiated, the resulting ESR spectrum contains five components (Fig. 9), of which the main copolymer radical probably is AA(H)–AN·. The radical concentrations at different levels of AN additions indicate that both AA(H)· and AA(T)· but probably not the allyl radicals AA(A)· react with the AN monomer (Fig. 10) (cf. Doi & Rånby, 1970). Plots of the conversion rates of M_1· to M_1–AN· radicals demonstrate the very high reactivity of VAc· radicals in comparison with the other monomer radicals in decreasing order AAc·, SVS·, AA·, SMAS· and MA· (Fig. 11). The slopes of the plots of conversion ratios versus AN monomer concentration are given in Table 2 together with reciprocal reactivity ratios $1/r_1$ for the monomer pairs and the Q_1 and e_1 values from the literature. The relative rates of conversion for the

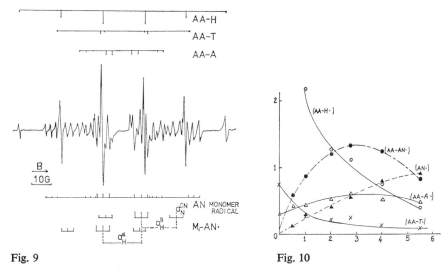

Fig. 9

Fig. 10

Fig. 9. ESR spectrum of a copolymerizing system of allyl alcohol [AA] = 0.055 mol l⁻¹ and acrylonitrile ([AN] = 0.020 mol l⁻¹) showing head (H), tail (T) and allyl radicals (A) of AA and AN monomer (AN·) and copolymer radicals (AA–AN·).

Fig. 10. *Abscissa:* [AN]/mol $l^{-1} \times 10^2$; *ordinate:* rad. conc. (arb. unit).

Radical concentrations measured from ESR spectra during copolymerization of AA with AN at different AN concentrations in the AA–AN system. [AA] = 0.055 mol l⁻¹. (AA–H·) head radical of allyl alcohol, (AA–T·) tail radical of AA, (AA–A·) allylic radical of AA, (AN·) monomer radical of acrylonitrile, (AA–AN·) a copolymer radical, probably of the structure $HO-CH_2-CH(CH_2OH)-CH_2-\dot{C}H(CN)$.

AAc·, SVS·, AA· and SMAS· monomer radicals are smaller than those expected from the $1/r_1$ values. MA· radicals show an especially low reactivity, which mainly is due to steric effects.

Copolymerization studies at pH 7 are listed in Table 3 and compared with data obtained at pH 1.4 in the previously described experiments. The radical reactivity of sulfonic acid monomers such as SVS and SMAS increases almost

Table 2. *The Slopes of the Plots of Conversion Ratios* $[M_1 - AN^{\cdot}] / [M_i]$ *versus* $[AN]$ *and the Reciprocal Monomer Reactivity Ratios* $1/r_1$ *Calculated from* Q_1 *and* e_1 *Values Given in the Literature*

Monomer Pair	Slope × 10⁻² of $[M_1 - AN^{\cdot}]/[AN^{\cdot}]$ versus [AN]	$1/r_1$	Q_1	e_1
VAc–AN	3.0	31.6	0.026	−0.22
AAc–AN	0.66	30.1	0.028	−1.13
SVS–AN	0.49	8.23	0.093	−0.02
AA–AN	0.41	8.86	0.052	−0.29
SMAS–AN	0.24	1.56	0.27	0.69
MA–AN	0.08	27.6	0.23	2.25

Q_2 (AN) = 0.60 e_2 (AN) = 1.20

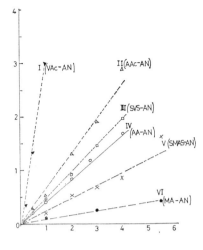

Fig. 11. *Abscissa:* $[AN]/mol\ l^{-1} \times 10^2$; *ordinate:* $[M_1-AN^{\cdot}]/[M_1^{\cdot}]$.

Plots of concentration ratios $[M_1-AN^{\cdot}]/[M_1^{\cdot}]$ versus concentration of AN for the systems I, VAc–AN (▼---▼); II, AAc–AN (△--△); III, SVS–AN (□-··-□); IV, AA–AN (○—○); V, SMAS–AN (×-··-×); VI, MA–AN (●---●). $[M_1] = 0.055\ mol\ l^{-1}$.

to the double with the increase in pH from 1.4 to 7.0, while the radical reactivity of AAc remains constant. At pH 7 the sulfonic acid group is dissociated and becomes an anion which gives the radical a negative character. This may explain the increase of the reactivity of the sulfonic acid radical with AN monomer which has a positive character.

Discussion

Using the main results from two copolymerization studies in aqueous solution, one with vinyl acetate and one with acrylonitrile as reference monomer, it is demonstrated how the ESR method can be applied for direct measurements of monomer and monomer radical reactivities in aqueous solution under different conditions of initiation, pH and monomer concentration.

These investigations are part of a research program on radical polymerization supported by The Swedish Board for Technical Development which is gratefully acknowledged.

Table 3. *The Reactivities of Monomer Radicals with AN at Different pH.* $[M_1] = 0.055$ *mol* l^{-1}, $[AN] = 0.020$ *mol* l^{-1}

M_1	$[M_1 - AN^{\cdot}]/[M_1^{\cdot}]$	
	pH 1.4	pH 7.0
AAc	1.30	1.41
AA	0.94	1.23
SVS	0.82	1.75
SMAS	0.54	1.06

References

Alfrey, Jr., T., Bohrer, J. J. & Mark, H., Copolymerization. Interscience, New York 1952.

Dixon, W. T. & Norman, R. O. C., J. Chem. Soc., *1963*, 3119.

Doi, Y. & Rånby, B., J. Polymer Sci. C, *31*, 231 (1970).

Fessenden, R. W. & Schuler, R. H., J. Chem. Phys., *39*, 2147 (1963).

Fischer, H., Z. Naturforsch., *19a*, 866 (1964).

Ham, G. E., (ed.) Copolymerization. Interscience, New York 1964.

Heller, C. & McConnell, J. Chem. Phys., *32*, 1535 (1960).

Izumi, Z. & Rånby, B., unpublished work (1972).

Takakura, K. & Rånby, B., J. Polymer Sci., B, *5*, 83 (1967).

Takakura, K. & Rånby, B., Adv. in Chemistry Series (ACS), No. *91*, Addition and Condensation Polymerization Processes, p. 125–144 (1969) *a*.

Takakura, K. & Rånby, B., J. Polymer Sci., C, *22*, 939 (1969*b*).

Takakura, K. & Rånby, B., J. Polymer Sci., A-1, *8*, 77 (1970).

Yoshida, H. & Rånby, B., J. Polymer Sci., C, *16*, 1333 (1967).

Discussion

Chapiro

The method described by Prof. Rånby is extremely powerful, because it makes it possible to directly measure radical reactivities towards monomers. Usually reactivity ratios in copolymerizations are derived from overall compositions of copolymers and this only gives a statistical picture of the system. Moreover, we have recently investigated a number of systems in which the monomer may become associated in regularly ordered aggregates (e.g. acrylic acid, acrylamide) and several monomer molecules may then add simultaneously via a "zip" propagation process (see paper presented at the Amer. Chem. Soc., Boston (1972), to be published in Adv. Chem. Series). Under such conditions reactivity ratios derived from overall compositions may be misleading and direct determination of monomer reactivities would be particularly pertinent.

Rånby

It should be observed, however, that radical polymerization reactions under "normal" conditions, e.g. as used in industrial processes, cannot be studied with the ESR method due to the low radical concentration. In the flow cell you have mixing and diffusion problems which may influence the rate measurements.

Sohma

I wish to make a comment on the applicability of the changing flow-rate method to determine the rate constant of the reaction.

The determination of the rate constant by changing flow is based on the assumption that mixing of two liquids is complete and independent of the flow rate. However, it is not experimentally verified.

In our laboratory the rate constants were determined by the two different methods: the changing flow-rate and the rapid scanning for the decay of the complex of Ti^{3+} and hydrogen peroxides by stopping flow. The different rate constants were obtained by the two different methods. The discrepancy between these experimental values is too large to be considered as an experimental error. This discrepancy indicates that mixing of the two liquids was not complete for the slow flow rate, and there exists a critical flow-rate below which the mixing is not complete. Thus, determination of the rate constant by the changing flow methods is not more reliable for the case of the slower flow than the critical flow rate which is dependent on the design of the mixing cell.

Rånby

The results presented by Prof. Sohma demonstrate that kinetic data on the $TiCl_3$–H_2O_2 radical-generating system by continuous-flow and stopped-flow procedures need not agree and he attributes this discrepancy of possible inadequacies of the continuous-flow method with respect to the efficiency of the rapid-mixing process, especially at low flow rates. The efficiency of the mixing of reactants is a well known limitation to the successful use of both the continuous-flow and the stopped-flow techniques in solving kinetic problems (e.g., Caldin, E. F., Fast Reactions in Solution, Blackwell, Oxford 1964). This efficiency naturally depends on the design of the mixing chamber employed. It would therefore be interesting to have a description of the design of the mixing chambers utilized in Prof. Sohma's investigations.

P. Smith

Professor Rånby's results seem to show that relative rate-constant data for short-chain radicals taken by way of the $TiCl_3$-H_2O_2 radical-generating method are in accord with the corresponding relative rate constant data taken by conventional methods which are for the analogous long-chain radicals. Would Prof. Rånby care to comment whether his sort of studies enable him to obtain evidence on how the absolute values of rate constants in radical copolymerization processes depend on chain length?

Rånby

In the ESR spectra we can usually only distinguish dimer, trimer etc. radical as a group from monomer radicals. Therefore, we cannot give direct evidence on variations in rate constants as a function of chain length. With the high rate of initiation necessary in the flow method, only oligomers can usually be expected in the reaction products.

INITIATION REACTION

High Energy Initation

ESR Study of γ-Irradiated Alcohols Adsorbed on Magnesium Oxide Powders

By D. R. Smith and A. J. Tench

Atomic Energy of Canada Limited, Chalk River Nuclear Laboratories, Chalk River, Ontario, Canada and Atomic Energy Research Establishment, Harwell, Berks, UK

Summary

ESR techniques are used to study the radiation chemistry of methanol and ethanol adsorbed on MgO powder. It is suggested that radical formation occurs by processes involving hydrogen abstraction by positive hole charge-carriers and by surface O^- centres in a range of environments, and dissociative electron attachment by reaction with surface trapped electrons. These data are correlated with literature data on similar processes obtained under well defined conditions.

Introduction

The study of the radiation chemistry of gas/solid heterogeneous systems is of interest because it offers the possibility of controlled modification of the electronic properties of solids to make new types of catalysts. Ionizing radiation is degraded within the solid to form charge-carriers such as electrons and positive holes, which in the case of a high surface area solid can reach the surface. The charge-carriers can be trapped at surface defect sites to form F- or V-type centres or they can recombine at impurities. Either way, the energy deposited by irradiation becomes available at the surface to initiate free radical or ionic chemical reactions such as polymerization. This work is relevant, in a fundamental way, to processes involved in surface polymerization reactions, whether induced by ionizing radiation or other means. Such research is also informative on how polymer coatings are affected by ionizing radiation and on radiation stability of "filled polymers". For example polymers may be used as seals, insulators and corrosion inhibitors in the nuclear industry and often contain fillers such as glass beads, metal or metal oxide to enhance radiation and thermal stability.

The aim of the work reported herein is to understand the fundamental chemistry of the system, metal oxide/adsorbed organic molecule, when exposed to ^{60}Co γ rays. The system, MgO/adsorbed alcohols, was chosen because of the considerable data available concerning surface species on

MgO and their interaction with alcohols (Smith & Tench, 1968, 1969; Tench et al., 1972). MgO is used in the relatively high surface area powder form to provide adsorption of sufficient alcohol in a monolayer to give measurable electron spin resonance signals, after irradiation. The data do not involve polymerization processes directly, but demonstrate some fundamental reactions which could be involved in the initiation processes.

Irradiation of MgO powder at 77 K in vacuo leads to an ESR signal associated with a V_s^- centre and a weak signal due to electrons trapped at the surface, F_s^+ centres (Eley & Zammitt, 1971; Tench, 1973; Nelson et al. 1967). In the presence of hydrogen, the V_s^- centre is eliminated and the $F_s^+(H)$ centres are greatly enhanced together with a chemisorption of hydrogen (Tench & Nelson, 1968).

Throughout this paper we have used a notation system proposed by Tench (1972, 1971), in which an F_s^+ centre is an electron trapped in a surface oxygen vacancy, the V_s^- centre is a positive hole trapped on one of the oxygen ions surrounding a surface cation vacancy and $(O^-)_s^+$ refers to an O^- ion in a normal O^{2-} ion site at the surface. In this notation, the subscript refers to the surface and the charge state refers to the overall effective charge state of the defect referred to the normal lattice charge. In this context, the V_s^- centre and $(O^-)_s^+$ are the same chemically but differ in their geometric environment.

The $F_s^+(H)$ centre is very reactive when exposed to gases or vapours. Thus exposure to CO_2 (Lundford & Jayne, 1965), O_2 (Tench & Holroyd, 1968), CH_3OH (Smith & Tench, 1968, 1969) or C_2H_5OH (Smith & Tench, 1969) leads to new trapped radicals CO_2^-, O_2^-, CH_2O^-, CH_3CHO^-, respectively. With N_2O (Tench et al., 1972; Tench & Lawson, 1970; Williamson et al., 1971) $(O^-)_s^+$ is formed and can be reacted with freshly admitted hydrogen to regenerate the $F_s^+(H)$ centre (Tench, Lawson & Kibblewhite, 1972; Tench & Lawson, 1970; Williamson, Lunsford & Naccache, 1971). $(O^-)_s^+$ reacts with O_2 (Tench, Lawson & Kibblewhite, 1972; Tench & Lawson, 1970; Naccache, 1971), CO (Tench, Lawson & Kibblewhite, 1972; Tench & Lawson, 1970; Nacchache, 1971) and CO_2 (Tench, Lawson & Kibblewhite, 1972; Tench & Lawson, 1970) to produce $(O_3^-)_s$, $(CO_2^-)_s$, and $(CO_3^-)_s$, respectively and its abstracts from methanol and other alcohols to form $(CH_2O^-)_s$, for example (Tench, Lawson & Kibblewhite, 1972). Admission of propene destroys $(O^-)_s^+$ with no new signal appearing (Tench, Lawson & Kibblewhite, 1972). Ethylene reacts with O^- to form a new radical, tentatively identified as $(CH_2CH_2O^-)_s$ (Naccache, 1971), but the unusually low proton hyperfine couplings, 4G, make this assignment questionable. Recent work with CD_2CD_2 suggests a cyclic radical of the form $(CD_2\text{--}CD_2^-)_s$ (Tench, unpubl.).

When MgO powder is exposed to gas phase hydrogen atoms the $F_s^+(H)$ centres (and their blue colour) are formed (Smith & Tench, 1968). For $F_s^+(H)$ centres to be trapped, oxygen ion vacancies have to be created during the radiation process. Simple gamma radiolysis of MgO in vacuo forms these with very low efficiency and the concentration of F_s^+ centre reflects the low vacancy concentration on the surface of the powder before irradiation. The high yield in the presence of H_2 implies the creation of vacancy traps and a possible mechanism for $F_s^+(H)$ centre formation involves reactions (1) to (3), since the initial concentration of cation vacancies is much higher than that of the anion vacancies on the unirradiated powder.

$$MgO \xrightarrow[\gamma\text{-rays}]{\text{in vacuo}} V_s^- + F_s^+ \text{ (small yield)} \tag{1}$$

$$V_s^- + H_2 \longrightarrow OH^- + H \tag{2}$$

$$H + O^{2-} \longrightarrow OH^- + F_s^+(H) \quad \text{i.e., a trapped electron} \tag{3}$$

These reactions demonstrate an important feature of this type of system. Initially, energy is predominantly deposited in the substrate leading to the formation of charge-carriers which migrate to the surface and are trapped to form V_s^- and F_s^+ centres. Subsequent chemistry involving the adsorbed molecules can be controlled by reactions with species or defects formed on the substrate surface. In this case charge-carriers appear to be involved directly as they are in the MgO/O_2 (Nelson et al., 1971) system. In other situations, for example in the azoethane/MgO (Rabe et al., 1966) system energy transfer appears to occur without directly involving charged species.

Another important feature is that, due to the nature of the trapping processes, diffusion is often greatly inhibited, and species which would react in milliseconds or less in the liquid phase can be stable for minutes and even days when isolated on a surface. The equivalent of migration can occur however if adsorbed parent molecules are sufficiently close by to permit chain transfer (via hydrogen transfer) (Smith & Tench, 1969). Charge migration or propagation of a polymer chain are other examples of how movement of a paramagnetic site could arise.

Experimental

Preparation of samples was described earlier (Smith & Tench, 1969), these being the blank samples prepared for that work. [60]Co γ irradiations were performed at 77 K for 2.5 h to a total dose of 1 Mrad. Prior to ESR measurement, the radiolytically induced signal in the silica tube was eliminated by flaming. During this manipulation the powder was maintained

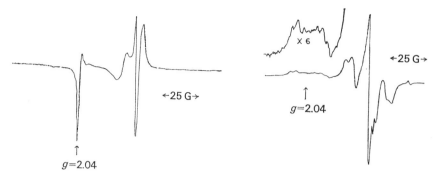

Fig. 1   Fig. 2

Fig. 1. Initial first-derivative ESR spectrum of pure MgO powder, γ irradiated at 77 K showing the $g = 2.04$ feature of V_s^-. Other lines arise from the saturated F_s^+ centre and an unknown defect probably involving O^- in the bulk.

Fig. 2. Initial first-derivative ESR spectrum of methanol adsorbed on MgO powder, γ irradiated at 77 K.

at 77 K in the other end of the tube. Transfer of powder from one end of the tube to the other was done under liquid nitrogen.

ESR spectra were obtained using a Varian V-4502 X-band spectrometer with 100 kHz field modulation. Unless otherwise specified, spectra are first derivative at 77 K with field increasing from left to right.

Results and Discussion

A typical ESR spectrum of γ-irradiated MgO powder, with no adsorbed alcohol is shown in Fig. 1. The important feature to note in this work is the low field line at $g = 2.04$. This is the $g_\perp$ feature of the V_s^- centre (Eley & Zammitt, 1971; Tench; Nelson, Tench & Harmsworth, 1967) and is similar to the $g_\perp$ for $(O^-)_s^+$ observed in MgO (Tench, Lawson & Kibblewhite, 1972, Tench & Lawson, 1970; Williamson, Lunsford & Naccache, 1971). Evidently there is some variation in trapping environment as the wings of this absorption are very broad. It will be seen in the following results that these sites are less reactive than those giving the narrower part of the absorption. This V_s^- centre is stable in vacuo at 77 K but slowly decays on warming to 300 K.

Data given below for spin concentrations and yields must be considered to be approximate in absolute terms, though relative to each other, and the concentration data in Smith & Tench, 1969; the precision is probably $\pm 10\%$.

Adsorbed Methanol

As in Smith & Tench, 1969, two coverages, 3×10^{-3} and 1.5×10^{-3} ml ROH g^{-1} MgO, were tried. The latter sample was pumped on for 30 min at room temperature after alcohol adsorption. The spectrum and intensity, Fig. 2, were the same in each case. The main spectrum at $g = 2.003$ is typical of a CH_2R type radical with axially symmetric hyperfine couplings $A_{||} = 23.6$ and $A_{\perp} = 13.2$ G($A_{average} = 16.7$). We know that CH_3OH adsorbs on MgO as $(CH_3O^-)_s$, giving an infrared spectrum which is not decreased by pumping (Tench, Giles & Kibblewhite, 1971). The proton hyperfine parameters being in reasonable agreement with those for $(CH_2O)^-Mg^{2+}$ (Smith & Tench, 1969) we identify this radical as CH_2O^-. The concentration of this radical in each case is about 1.5×10^{17} spins g^{-1}, about 3.4 times greater than that obtained by reaction with a gaseous stream of H atoms (Smith & Tench, 1969). The room temperature decay data in this work are too crude to establish kinetic order, especially since the concentration-time curves increased initially. The $(CH_2O^-)_s$ concentration in the sample containing 3×10^{-3} ml ROH g^{-1} MgO increased by 42% in 15 s before decaying with an initial half life of 7 min while that in the lower coverage sample increased by 26% in 45 s before decaying with an initial half life of 11 min. It may have increased more in a shorter time. These intensity vs time measurements were at 77 K after warming for a measured time to room temperature.

The initial $(CH_2O^-)_s$ concentration, about 1.5×10^{17} spins g^{-1} corresponds to a radiation chemical yield $G \simeq 0.2$. (G = number of molecules, etc. formed per 100 eV energy deposited in the sample.)

There are other prominent features in the spectrum. At high field, $g = 1.9796$, there is a weak absorption due to Cr^{3+} impurity (higher concentration than in the MgO used for Fig. 1). At $g = 2.006$, superimposed on the central line of the $(CH_2O^-)_s$ spectrum, there is a weak unidentified line. Most of the surface V_s^- centres have ben eliminated by the presence of methanol; however there is a broad residual adsorption at $g = 2.04$ (Fig. 2), which corresponds to the $g_{\perp}$ feature of V_s. An important observation is that 50–75% of this signal disappears at room temperature during the initial seconds while the concentration of $(CH_2O^-)_s$ is increasing. Combining this with the observation of Tench et al. (1972) that exposure to CH_3OH of surface $(O^-)_s^+$ centres produced from N_2O destroys the $(O^-)_s^+$ ion and forms $(CH_2O^-)_s$, we conclude that the same reaction is occurring here, since the difference between the V_s^- centre and the $(O^-)_s^+$ is geometric and the chemical identity should be essentially the same. Evidently the V_s^- centres have a range of reactivities; those absorbing near the centre of the ESR line can react with $(CH_3O^-)_s$ even at 77 K, while those contributing to the broad underlying absorption react at room temperature.

The irradiated CH_3OH/MgO samples were colourless, with no evidence of the blue colour or the narrow ESR line at $g = 1.9998$ due to surface trapped electrons near a hydroxyl group ($F_s^+(H)$ centres) (Tench & Nelson, 1968).

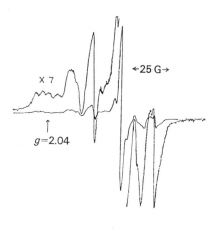

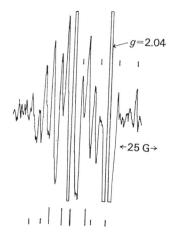

Fig. 3   Fig. 4

Fig. 3. Initial first-derivative ESR spectrum of ethanol adsorbed on MgO powder, γ irradiated at 77 K.

Fig. 4. Final ESR spectrum, second-derivative at room temperature of ethanol adsorbed on MgO powder, γ irradiated at 77 K and warmed to room temperature for several hours. The stick spectra indicate the two quartets. One line is unusually strong due to residual adsorption by the $F_s^+(H)$ centre. The arrow indicates the Cr^{3+} singlet, and the four satellite hyperfine lines from ^{53}Cr ($I=3/2$, natural abundance $= 9.54\%$) are also shown.

The weak line at $g=2.0006$ might be due to a low concentration of F_s^+ centres (Tench & Nelson, 1968).

Adsorbed Ethanol

Samples containing 3×10^{-3} and 1.5×10^{-3} ml C_2H_5OH g^{-1} MgO were studied. The results were essentially the same in each case. Fig. 3 shows the initial spectra. The quintet of broad lines, $g=2.003$, $A_H \simeq 21$–22 G is typical of the α-ethanol type of radical. Here we suspect it is a mixture of CH_3CHOH and CH_3CHO^- as will become more evident below. The total concentration is about 2.5×10^{17} spins g^{-1} corresponding to $G \simeq 0.4$. A weak Cr^{3+} line is seen superimposed on the high field outer line of the quintet and the broad residue of the V_s^- line at $g=2.04$ is seen partially superimposed on the low field outer line of the quintet. As in the methanol/MgO system, most of this species has been eliminated by the presence of ethanol.

In this system the irradiated sample is coloured sky-blue and, superimposed on the centre line of the ethanol quintet is the typical, easily saturated surface trapped electron ($F_s^+(H)$) line at $g=1.999$ (Tench & Nelson, 1968). Samples were warmed to room temperature for short intervals, recooled, and the ESR spectrum studied again at 77 K. The intensity of the $F_s^+(H)$ line decreased rapidly as did the overall intensity of the quintet.

However, there were slight changes in line shapes and relative intensities in the quintet. The $F_s^+(H)$ line was 75% gone in 6 min and 90% gone after an additional 7 min at room temperature with a concurrent decrease in the blue colour intensity. During the same period the spectral details of the quintet became constant after several minutes and the intensity of the quintet diminished by about 75%. The V_s^- signal also diminished considerably in the first few minutes.

This left the room temperature narrow line spectrum shown in Fig. 4. This is due to CH_3CHO^-, the same spectrum as was formed by exposing $F_s^+(H)$ centres to ethanol vapour (Smith & Tench, 1969) It is a spectrum of two quartets with most of the lines resolved and the proton hyperfine couplings are $A_\alpha = 13.5$ G, $A_{CH_3} = 20.6$ G. This radical on the surface of MgO is also recognized by its great stability (Smith & Tench, 1969). In this work, on standing at room temperature, about half the $(CH_3CHO^-)_s$ radicals decayed in 2 h but the remaining fraction was stable for several days.

We feel that initially the quintet spectrum involved predominantly CH_3CHOH radicals formed from physisorbed CH_3CH_2OH. We know that it decays in several minutes at room temperature (Smith & Tench, 1969). The changes in detail of the quintet noted above are caused by the concurrent formation of $(CH_3CHO^-)_s$ via reaction with $F_s^+(H)$ centres. The somewhat longer half life will be an artifact caused by this and by: (*a*) any stable CH_3CHO^- fraction which existed initially as a result of reaction between V_s^- and chemisorbed $CH_3CH_2O^-$; (*b*) the concurrent formation of more CH_3CHOH via reaction between CH_3CH_2OH and the residual V_s^- centres.

Mechanism and Conclusions

As in all heterogeneous systems the phenomena involved are complex. In this case, we know some details of the adsorbed species before irradiation and infrared work has shown that alcohol molecules can be present in at least two forms. Some may be simply physisorbed RCH_2OH but others are chemisorbed, bound to Mg^{2+} as $(RCH_2O^-)_s$ (Tench, Giles & Kibblewhite, 1971). These adsorbed species can then react with charge-carriers and excitons available at the surface during irradiation in competition with existing anion or cation vacancy traps. On warming, a further reaction can take place and the alcohol molecules can diffuse to and react with V_s^- and F_s^+ centres.

The most important observation is that in the presence of alcohols the ESR signals from the V_s^- centre are destroyed. This suggests that hydrogen can be abstracted from the adsorbed species. It seems likely that charge-carriers such as the positive hole arising at the surface can react directly with the adsorbed alcohol. This could give a charged $(CH_3OH)_s^+$ species as suggested (Zhabrova et al., 1966; 1967) for alcohols adsorbed on silica; subsequent charge neutralization by capture of an electron was assumed to lead to the loss of a hydrogen atom. On the other hand we have substantial evidence (Smith & Tench, 1968; 1969; Tench, Lawson & Kibblewhite, 1972) on MgO that the chemical entity O^- will react with alcohols directly, and, writing O^- to mean either the hole carrier or the trapped hole and similarly for e^-, then the reactions (4) to (7) below could be expected to occur at the surface either initially or during warming up.

No evidence is found for formation of OH via the proton transfer reaction (8), but OH would abstract via (9) just as easily as O^- and we would not then see anything but removal of O^- and formation of alcohol radicals. We cannot really distinguish between (8)+(9) vs (4) in this work.

$$O^- + \frac{RCH_2OH}{RCH_2O^-} \longrightarrow \frac{RCHOH}{RCHO^-} + OH^- \tag{4}$$

$$e^- + RCH_2OH \longrightarrow RCHO^- + H_2 \tag{5}$$

$$\text{or}$$

$$RCH_2O^- + H$$

$$H + \frac{RCH_2OH}{RCH_2O^-} \longrightarrow \frac{RCHOH}{RCHO^-} + H_2 \tag{6}$$

$$2H \longrightarrow H_2 \tag{7}$$

$$O^- + ROH \longrightarrow OH + RO^- \tag{8}$$

$$OH + \frac{RCH_2OH}{RCH_2O^-} \longrightarrow \frac{RCHOH}{RCHO^-} + H_2O \tag{9}$$

These reactions leave a residue of unreacted V_s^- at 77 K and also $F_s^+(H)$ in the C_2H_5OH/MgO system. On warming up to room temperature reaction (4) occurs, to further reduce the V_s^- concentration and form more alcohol radicals. The surface trapped electrons $F_s^+(H)$ also undergo a reaction analogous to (5).

There are subtle features in this scheme, relating to final trapping sites. $RCHO^-$ formed in (4) will probably be bonded to Mg^{2+}. However, $RCHO^-$ formed by reaction with $F_s^+(H)$ analogous to (5) will likely be trapped at the surface oxygen ion vacancy which formerly trapped the $F_s^+(H)$ centre. Though these two different trapping sites for $RCHO^-$ may result in different stabilities this work would not distinguish between them. It seems likely the nature of these trapping sites accounts for the far greater stability of $RCHO^-$ compared to RCHOH (lifetime of days compared to minutes) as the latter would be relatively mobile.

This work was carried out during the attachment of D. R. Smith from Atomic Energy of Canada Limited, Chalk River, to the Atomic Energy Research Establishment at Harwell during 1967–1968.

D. R. Smith wishes to acknowledge the support and encouragement of Dr C. B. Amphlett at Harwell.

References

Smith, D. R. & Tench, A. J., Chem. Comm., *18*, 1113 (1968).

Smith, D. R. & Tench, A. J., Can. J. Chem., *47*, 1381 (1969).

Tench, A. J., Lawson, T. & Kibblewhite, J. F. J., J. Chem. Soc., Faraday. In press (1972).

Eley, D. D. & Zammitt, M. A., J. Catalysis, *21*, 377 (1971).

Tench, A. J., Unpublished work.

Nelson, R. L., Tench, A. J. & Harmsworth, B. J., Trans. Faraday Soc., *63*, 1427 (1967).

Tench, A. J. & Nelson, R. L., J. Colloid and Interface Sci., *26*, 364 (1968).

Tench, A. J., 1971, AERE R 6888.

Lunsford, J. H. & Jayne, J. P., J. Phys. Chem., *69*, 2182 (1965).

Tench, A. J. & Holroyd, P., Chem. Comm., *18*, 471 (1968).

Tench, A. J. & Lawson, T., Chem. Phys. Letters, *7*, 459 (1970).

Williamson, W. B., Lunsford, J. H. & Naccache, C., Chem. Phys. Letters, *9*, 33 (1971).

Naccache, C., Chem. Phys. Letters, *11*, 323 (1971).

Tench, A. J., Unpublished work.

Nelson, R. L., Hale, J. & Harmsworth, B. J., Trans. Faraday Soc., *67*, 1164 (1971).

Rabe, J. G., Rabe, B. & Allen, A. O., J. Phys. Chem., *70*, 1098., (1966).

Tench, A. J., Giles, D. & Kibblewhite, J. F. J., Trans. Faraday Soc., *67*, 854 (1971).

Zhabrova, G. M., Vladimirova, V. I., Kadenatsi, B. M., Kazanskii, V. B. & Pariiskii, G. B., J. Catalysis, *6*, 411 (1966).

Vladimirova, V. I., Zhabrova, G. M., Kadenatsi, B. M., Kazanskii, V. B. & Pariiskii, G. B., Dokl. Adad. Nauk. SSSR, *164*, 361 (1965).

Zhabrova, G. M., Vladimirova, V. I., Kadenatsi, B. M., Kazanskii, V. B. & Pariiskii, G. B., Zhurn. Fiz. Khim., *61*, 1898 (1967).

Discussion

Yoshida

In the reaction of electrons with ethanol do you know whether the molecules or H atoms are formed initially?

D. R. Smith

I believe H_2 molecules are directly formed because H atoms would react with the matrix to form more electrons

$$H + O^{2-} \rightarrow e^- + OH^-$$

and we would have a chain reaction. Furthermore, Bennett & Mile photolytically induced the reaction of e⁻ with ethanol in ethanol solid containing benzene. Only CH₃ĊHO⁻ radicals were formed and no C₆H₇· was formed. This means that no free H atoms were formed and H₂ must be formed in one step.

Williams

I notice that the radicals you identified in the MgO methanol system did not include the methoxy radical, CH₃O·. In recent work on the radiolysis of liquid methanol in the presence of a spin trap, we observed the adduct corresponding to the addition of this radical and this was the major product at low temperature. I wonder if this radical may be formed in your system but is difficult to detect in the solid state.

D. R. Smith

We do not see this, but the hyperfine couplings would be low and it is possible that some CH₃O· radicals are present, with the spectrum masked by the ĊH₂O⁻ spectrum.

Ivin

My colleagues, R. C. Pink et al. have formed paramagnetic species simply by adding certain compounds to alumino-silicates. Can this be done with MgO?

D. R. Smith

I believe that Tench & Nelson reported formation of TCNE⁻ on adding tetracyanoethylene to MgO.

Kinell

Do you have some evidence for any rotational motion of the alcohol radicals on the MgO surface? We have found that radicals or groups in radicals are rotating on the surface of silica gel even at 77 K.

Tsuji

Do you not get any information on the orientation of alcohol radicals on the surface of MgO from their ESR spectrum?

D. R. Smith

I will answer Prof. Kinell and Prof. Tsuji together.

   This information is mentioned in the text and in our earlier work (Smith & Tench, 1969) but it is useful to summarize it. The line shapes of the ĊH₂O⁻ spectrum show axial symmetry and we think this radical is rotating about the C–O bond, the rotation axis lying in a plane perpendicular to the surface. The radical ĊH₂O⁻ must be incapable of diffusion as it must be bonded to Mg^{2+}. However, the free radical site appears to be able to migrate via hydrogen transfer (Smith & Tench, 1969). The decay must be due to eventual combination with other $\dot{C}H_2O^-$ radicals or perhaps with V_s^- centres in this work.

Due to overlap with V_s^- or Cr^{3+} lines we were unable to analyze the line shapes in the ethanol radical spectra so we can't comment on rotational movement. $CH_3\dot{C}HOH$ must be relatively mobile and this is confirmed by its decay in a minute or two at room temperature. $CH_3\dot{C}HO^-$ will be unable to diffuse as it will be bonded to Mg^{2+} or the O^- portion will reside in an O^- vacancy (because the electron which reacted to form this radical was trapped there) and this accounts for the lifetime of days at room temperature.

Smith

It would be interesting to make $(O^-)_s^+$ or $F_s^+(H)$ on the surface of MgO powder and expose them to polymerizable monomers. $F_s^+(H)$ might initiate anionic polymerization. $(O^-)_s^+$ might initiate free radical polymerization by abstraction or addition and it might even be capable of initiating cationic polymerization by charge transfer.

Szwarc

In the case of anionic polymerization, the growing polymer chain, with the negative charge at the end, would tend to move this charge away from the surface. The development of space charge effects might hinder the polymerization.

ESR Study of Some Radiation-Induced Vinyl Radicals

By Rose Marx, Serge Fenistein and Lydia Bonazzola

Laboratoire de Physico-Chimie des Rayonnements de l'Université de Paris-Sud, Centre d'Orsay, 91-Orsay, France

Apart from the role of vinyl radicals in radiation-induced polymerization, their electronic structure may be of some interest.

It is well known that ESR spectroscopy may give very useful information on this problem, especially when ^{13}C couplings are available. The information obtained via α or β proton couplings are in general less easy to interpret. However in vinyl radicals the β methylene proton couplings are simply related to the π or σ character of the unpaired electron: the two couplings are equal in a π radical and unequal in a σ radical, giving a 3 line ESR spectrum in the first case and a 4 line spectrum in the second case.

Our previous results (Bonazzola, Fenistein & Marx, 1971) indicated that the electronic structure of α-substituted vinyl radicals may be closely related to the nature of the substituent.

We would like to report here some more evidence on this relation.

Saturated Substituents

Vinyl, $CH_2=\dot{C}H$ and methyl vinyl, $CH_2=\dot{C}CH_3$ radicals, have first been observed by Fessenden & Schuler (1963) in electron irradiated liquid ethylene and propylene at low temperature. The unequal coupling of the two β methylene protons have been attributed to a bent configuration of the radical, the unpaired electron being located in a σ type molecular orbital:

$$\tag{1}$$

$$\tag{2}$$

Radical (1) has also been observed in solid matrices at 4.2 K, the couplings of β protons still unequal in argon (Cochran et al., 1964) were equal in a neon matrix (Kasai et al., 1967). As a linear π configuration of this radical seems

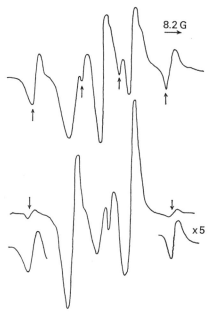

Fig. 1. *a*, ESR spectrum of γ-irradiated $CH_2 = \overset{\cdot}{C} - C(CH_3)_3$ four lines of *t*-butylvinyl radical (↑);

b, ESR spectrum of γ-irradiated $CH_2 = \overset{\cdot}{C} - C\overset{\displaystyle O}{\underset{\displaystyle H}{\diagdown}}$ two external lines of formylvinyl radical (↓)

8.2 G.

to be unlikely on the basis of the previous experimental results and theoretical calculation (Fessenden, 1967; Drago et al., 1967; Millie et al., 1968 etc.). This may be explained by a possible rapid reorientation of the radical in neon matrices.

Our observation of vinyl and methyl vinyl radicals produced by γ irradiation of CH_2CHCl and $CH_2CCl-CH_3$ adsorbed on molecular sieves are consistent with the coupling constants given by Fessenden (1963).

The spectrum of *t*-butyl vinyl radical:

$$\underset{H}{\overset{H}{\diagdown}} C = \overset{\cdot}{C} \overset{C(CH_3)_3}{\diagup} \qquad (3)$$

has been obtained in γ-irradiated dimethyl-3,3-butyne adsorbed on 13 X molecular sieves.

As shown on Fig. 1 *a* the 4 line spectrum of radical (3) is superimposed on a 3 line spectrum due to dimethyl-2,2-butyne-3-yl-radical:

$$H-C \equiv C \overset{\overset{\displaystyle CH_3}{|}}{\underset{\underset{\displaystyle CH_3}{|}}{-\overset{\cdot}{C}H_2}} \qquad (4)$$

Table 1

Radical	$a_\beta^H(CH_2)$ exp G	$\theta°$	E_{AU}	$a_\beta^H(CH_2)$ INDO calc. G
$\mathrm{H_2C{=}\dot{C}H}$ (with angle θ)	$35.2-68.5^a$ $37\ -65^b$	10 20 30 60	-15.557 -15.559 -15.562 -15.551	$53.64-65.77$ $45.24-67.17$ $37.37-66.07$ $21.13-57.77$
$\mathrm{H_2C{=}\dot{C}{-}CH_3}$	$32.9-57.9^b$	20 30 45	-24.053 -24.051 -24.041	$40.4-60$ $33.7-59.9$ $24.7-56.36$
$\mathrm{H_2C{=}\dot{C}{-}C(CH_3)_3}$	$33.6-59.6$	0 30 45	-49.274 -49.273 -49.260	$54.5-54.5$ $30.5-60.2$ $20.4-54.7$

[a] Cochran, 1964.
[b] Fessenden, 1963.

A computer program was used to deduce the coupling constants from the experimental spectrum. The values are reported in Table 1. One can see that the β-methylene couplings of tertbutyl vinyl are very close to those of methyl vinyl so that a bent σ structure may also be inferred.

To confirm this assumption an INDO calculation of total energy and spin density distribution as a function of the bending angle θ has been made for the vinyl, methyl vinyl and tertbutyl vinyl radical. The method and bond lengths used in this calculation are taken from Pople (Pople & Beveridge, 1968, 1970) and the results are shown in Table 1.

The best fit between experimental and calculated couplings and the configuration of lowest total energy do not occur for the same bending angle. But the energy barrier corresponding to the best fit ($\theta \simeq 25°$) is low enough to be neglected. This slight discrepancy between energetics and spin density distribution seems to be a rather general result of INDO calculations.

So there are now quite good experimental and theoretical evidence that vinyl radicals with a saturated alkyl α substituent have a π bent structure with a bending angle θ of about $30 \pm 5°$.

Unsaturated Substituents

Since the first π type vinyl radical, $CH_2{=}\dot{C}{-}CN$, was identified (Marx et al., 1967, 1969) some other α-substituted vinyl radicals have been studied:

$$CH_2{=}\dot{C}{-}C{\equiv}CH \text{ (Kasai et al., 1968)} \tag{5}$$

$$CH_2{=}\dot{C}{-}C_6H_5 \text{ (Bennet et al., 1971; Bonazzola et al., 1971)} \tag{6}$$

$$CH_2=\overset{\cdot}{C}-COOH \tag{7}$$

$$CH_2=\overset{\cdot}{C}-COOCH_3 \text{ (Bonazzola et al., 1971)} \tag{8}$$

Because of the great reactivity of vinyl radicals and of the parent molecules:

$$\underset{H}{\overset{H}{\diagdown}}C=C\underset{Y}{\overset{Cl}{\diagup}} \quad \text{or } HC\equiv C-Y$$

we obtained in most cases a mixture of radicals and the ESR spectra were very difficult to analyze unambiguously. In these cases an INDO calculation of the energy and spin density distribution was used to provide additional information on the most favourable configuration (Bonazzola et al., 1971). The best fit between experimental and calculated coupling constants was obtained for a π-type structure where the α carbon has an sp hybridization and the β-methylene protons lie in the antinodal plane of the unpaired electron.

We suggested that this structure may be related to the possibility for the unpaired electron to delocalize over the π orbitals of the substituent. A new result can be added to confirm this assumption. Formyl vinyl radical:

$$CH_2=\overset{\cdot}{C}-C\overset{\displaystyle O}{\underset{H}{\diagup}} \tag{9}$$

has been produced by γ irradiation of $CH=C-C\overset{O}{\underset{H}{\diagup}}$ adsorbed on 10 X

molecular sieve. The ESR spectrum shown on Fig. 1b is rather complicated, however the big doublet, split by 39 G, may be attributed to the growing radical:

$$\sim CH=C-CH=C-CH=\overset{\cdot}{C}-C\overset{\displaystyle O}{\underset{H}{\diagup}} \tag{10}$$
$$\quad\;\; |\qquad\quad\; | $$
$$\quad\;\; COH \quad\; COH$$

and the doublet split by 99.2 G to the external lines of formyl vinyl radical (9).

The central part of the spectrum remains unexplained and it is impossible to say if there is one or two more lines corresponding to the vinyl radical. An INDO calculation has been made in the same way as described previously (Bonazzola et al., 1971) the results given in Table 2 have been obtained with

$-C\overset{\displaystyle O}{\underset{H}{\diagup}}$ plane perpendicular to CH_2 plane. The bond length are those given by

Table 2

$\begin{array}{c}\text{H}\\ \diagdown\\ \text{C}=\dot{\text{C}}-\text{C}\diagup^{\text{O}}\\ \diagup \qquad\qquad \diagdown\\ \text{H} \qquad\qquad \text{H}\end{array}$	C—C Å	θ	E_{AU}	$\sum a_\beta^{\text{H}}(\text{CH}_2)$ INDO calc.	$a_\beta^{\text{H}}(\text{COH})$
$\sum a_\beta^{\text{H}}(\text{CH}_2)$ exp = 99.2 G	1.4 − 1.4	0°	− 39.97	$\left.\begin{array}{l}40\\40\end{array}\right\}$ 80	6.19
$a_\beta^{\text{H}}(\text{COH})$ exp ⩽ 4 G.		30°	− 49.96	$\left.\begin{array}{l}31.8\\46.6\end{array}\right\}$ 78.4	12.11
	1.35 − 1.35	0°	− 39.99	$\left.\begin{array}{l}42\\42\end{array}\right\}$ 84	4.96
		30°	− 39.98	$\left.\begin{array}{l}34.6\\49.6\end{array}\right\}$ 83	12.37

Pople (CH: 1.08 Å; C=O: 1.21 Å; C—C: 1.4 or 1.35 Å).[1] With $\text{C}\diagup^{\text{O}}_{\diagdown\text{H}}$ and

CH$_2$ in the same plane one obtains a very bad fit between experimental and calculated couplings and energy barriers are of the order of 20 kcal.

Finally it is interesting to check if the semi empirical relation

$$a_\beta = \varrho Q_\beta \qquad\qquad [1]$$

we proposed to calculate the methylene protons coupling in π vinyl radicals is valid for our formyl vinyl radical. $Q_\beta = 58.6$ G is the same as in alkyl radicals. The spin density ϱ is calculated using relation:

$$\varrho = 1(1 - \Delta\varrho_y)(1 - \Delta\varrho_{\text{CH}_2}) \qquad\qquad [2]$$

$\Delta\varrho_{\text{CH}_2} = 0.03$ is the spin density removed by the β-methylene group (Bonazzola et al., 1971); $\Delta\varrho_y$ is the spin density removed by the substituent (here the formyl group $-\text{C}\diagup^{\text{O}}_{\diagdown\text{H}}$).

$\Delta\varrho_{\text{COH}}$ may be calculated from the methyl proton coupling of the corresponding alkyl radical, $\text{CH}_3\dot{\text{CH}}-\text{C}\diagup^{\text{O}}_{\diagdown\text{H}}$. Unfortunately this radical has not been observed so we must use the most similar radical available in the literature:

[1] *Note added in proof:*
The best fit with experimental couplings has been obtained for $C_1 - C_2 = 1.35$ Å, $C_2 - C_3 = 1.4$ Å and $\theta = 0$ giving: $\Sigma a_{\text{H}}^\beta = 94.12$ G, $a_{\text{H}}^\beta(\text{COH}) = 5.24$ G, $E_{\text{AU}} = -39.99$.

Table 3

Radical	$a_\beta^H(CH_2)$ expt. G	$a_\beta^H(CH_2)$ INDO calc.	$a_\beta^H(CH_2) = \rho Q_\beta$ semi-empirical
$CH_2 = \dot{C}-CN$	48.2	50.2	48.6
$CH_2 = \dot{C}-C_6H_5$	41.5	46.6	41.3
$CH_2 = \dot{C}-COOH$	52.4	47.2	52.6
$CH_2 = \dot{C}-C \equiv CH$	43.4		44.5
$CH_2 = \dot{C}-COH$	49.6		49.4
$CH_2 = \dot{C}-H$	37–65	37.4–66.1 $(\theta = 30°)$	
$CH_2 = \dot{C}-CH_3$	33.9–57.9	33.7–59.9 $(\theta = 30°)$	
$CH_2 = \dot{C}-C(CH_3)_3$	33.6–59.6	34–59 $(\theta = 25°)$	

$$CH_2OH—CH_2—\dot{C}H—C{\overset{\displaystyle O}{\underset{\displaystyle H}{\Big\langle}}} \ .$$

Using the experimental coupling given by Corvaja & Fisher (1965) one finds $\Delta\rho_{COCH_3} = 0.131$. If we assume $\Delta\rho_{COH} \simeq \Delta\rho_{COCH_3}$ then $a(calc) = 58.6 \times 0.869 \times 0.97 = 49.4$ G which is in excellent agreement with the experimental coupling. So we have another example of α substituted vinyl radical which has a linear π structure and an unsaturated substituent allowing a delocalization of the unpaired electron.

In conclusion there seems to be, till now, no exception to the rule that α substituted vinyl radicals have a σ structure when the unpaired electron cannot delocalize on the substituent and a π structure when this delocalization is possible (Table 3).

References

Bennet, J. E. & Howard, J. A., Chem. Phys. Letters, 9, 460 (1971).
Bonazzola, L., Fenistein, S. & Marx, R., Mol. Phys., 22, 689 (1971).
Corvaja, C., Fisher, H. & Giacometti, G., Z. Phys. Chem. NF, 45, 1 (1965).
Cochran, E., Adrian, F. J. & Bowers, V. A., J. Chem. Phys., 40, 213 (1964).
Drago, R. S. & Petersen, H., J. Amer. Chem. Soc., 89, 5774 (1967).
Fessenden, R. Q. & Schuler, R. M., J. Chem. Phys., 39, 2147 (1963).
Fessenden, R. Q., J. Phys. Chem., 71, 74 (1967).
Kasai, P. H. & Wipple, E. B., J. Amer. Chem. Soc., 89, 1033 (1967).
Kasai, P. H., Skattebol, L. & Wipple, E. B., J. Amer. Chem. Soc., 90, 4509 (1968).
Marx, R. & Fenistein, S., J. Chim. Phys., 64, 1424 (1967).
Marx, R., Fenistein, S., Moreau, C. & Serre, J., Theor. Chim. Acta, 14, 339 (1969).
Millie, P. & Berthier, G., Int. J. Quant. Chem., 25, 67 (1968).

Formation and Decay of Radical Pairs in Vinyl Monomer Single Crystals

By Tomas Gillbro and Per-Olof Kinell[1]

The Swedish Research Councils' Laboratory, Studsvik, Nyköping, Sweden

In studies of polymerization in the solid state, the ESR technique has proved of great value in a large number of investigations. When single crystals are used, in particular, information can be obtained about the structure of radicals which initiate polymerization and propagating polymer radicals (see for example Adler & Petropoulos, 1965; Bamford, Eastmond & Sakai, 1963; O'Donnell, McGarvey & Morawetz, 1964; Ueda, 1964; Shioji, Ohnishi & Nitta, 1963). This method is advantageous when the monomers are easily crystallized at room temperature or slightly below room temperature (e.g., acids such as acrylic acid or acid salts such as barium methacrylate dihydrate). However, most monomers of interest are liquid at room temperature and only form single crystals with difficulty, and consequently only a few studies of these monomers have been published (Gillbro, Kinell & Lund, 1971). In the publication by Gillbro, Kinell & Lund (1971) radical pairs of monomer radicals were reported, and the present article can be regarded as an extension of the latter investigation. The monomers used in this work were methyl acrylate and methacrylonitrile. Radical pairs have previously been studied in a number of single crystals and some polycrystalline hydrocarbons (for a review see Lebedev, 1969).

The spin Hamiltonian of the radical pair can be written as follows, provided that the exchange of the two unpaired electrons is rapid: (Kurita, 1964; Iwasaki, Minakata & Toriyama, 1971).

$$\mathcal{H} = \beta \bar{g}\, \boldsymbol{BS} + (g_0 \beta/2)\, \boldsymbol{S} \times \sum_i (A_i^{\mathrm{I}} I_i^{\mathrm{I}} + A_i^{\mathrm{II}} I_i^{\mathrm{II}}) + (g_0^2 \beta^2/r^3) \times [\boldsymbol{S}^{\mathrm{I}} \boldsymbol{S}^{\mathrm{II}} - 3(\boldsymbol{S}^{\mathrm{I}} \boldsymbol{r})$$

$$\times (\boldsymbol{S}^{\mathrm{II}} \boldsymbol{r})/r^2] - B \sum_i (g_{ni}^{\mathrm{I}} \beta_{ni}^{\mathrm{I}} I_i^{\mathrm{I}} + g_{ni}^{\mathrm{II}} \beta_{ni}^{\mathrm{II}} I_i^{\mathrm{II}}) + J \boldsymbol{S}^{\mathrm{I}} \boldsymbol{S}^{\mathrm{II}} \tag{1}$$

here $\bar{g} = (g^{\mathrm{I}} + g^{\mathrm{II}})/2$ and $\boldsymbol{S} = (\boldsymbol{S}^{\mathrm{I}} + \boldsymbol{S}^{\mathrm{II}})$. I and II denote the paired radicals, i is the ith coupling nucleus in each paired radical, and r is the vector connecting the two unpaired electrons where r is taken to be the average distance between them. If B is larger than β/r^3 the resonance field strength for the $\Delta M_s = 1$ transition is as follows, neglecting the nuclear Zeeman term in eq. (1):

$$\mathcal{H} = B_0 \pm 1/2\, d - 1/2 \sum_i (A_i^{\mathrm{I}} m_i^{\mathrm{I}} + A_i^{\mathrm{II}} m_i^{\mathrm{II}}) \tag{2}$$

[1] Present address: Department of Physical Chemistry, University of Umeå, S-901 87, Umeå, Sweden.

where

$$d = \frac{3g\beta_0}{2r^3}(1 - 3\cos^2\theta) \tag{3}$$

In eq. (3) θ is the angle between $\boldsymbol{B}$ and $\boldsymbol{r}$. Eqs. (2) and (3) exhibit two of the main peculiarities of the radical pair, namely the anisotropic doublet spin–spin coupling and the hfs coupling which is only half as large due to the spin delocalization over two radicals. The third characteristic of radical pairs is the weak $\Delta M_s = 2$ transition, corresponding to the simultaneous flipping of two electron spins. The resonance magnetic field for the $\Delta M_s = 2$ transition is

$$\mathcal{H} = B_0/2 - 1/2 \sum_i (A_i^I m_i^I + A_i^{II} m_i^{II}).$$

The intensity of the $\Delta M_s = 2$ transition is normally about 1 000 times weaker than the intensity of the $\Delta M_s = 1$ transition (Iwasaki, Ichikawa & Ohmori, 1969).

Although the study of radical pairs in itself is of great interest for the spectroscopist, nevertheless for present purposes the information that can be gained about the fundamental processes involved in radiation-induced radical formation is of greater interest.

One big problem still to be solved in the radiation chemistry of vinyl monomers concerns the nature of the primary steps in the production of hydrogen-addition type vinyl radicals. It is generally assumed that a free hydrogen atom adds on to the double bond of a neutral monomer molecule. However, there is no definite evidence for such a mechanism so long as no free hydrogen atoms have been detected in monomer systems by ESR or other techniques. In this paper the possibility will be considered that hydrogen-addition type radicals are produced in the monomer crystal by a radiation-induced ionization of neighbouring molecules in the crystal matrix, followed by a proton transfer process. The stability of the radical pairs has also been studied.

Experimental

The monomers, methylacrylate (MA) and methacrylonitrile (MAN), were of "purum" grade and stabilized. Before use the monomers were freed from the stabilizer by distillation under vacuum twice. They were then dried over heated molecular sieves and stored in a refrigerator until used. The purified monomer was filled into a Suprasil quartz tube, one end of which was elongated into a thin capillary. The samples were degassed several times before they were sealed off at about 10^{-4} torr and 77 K. The crystals were prepared by a freezing technique (Dahlgren et al., 1971). The capillary end of the tube was lowered into a cooling bath at a speed of about 3 mm/h and the temperature of the bath was kept 5–10°C below the melting point of the monomer (for MA the melting point is −81°C and for MAN it is −36°C). The crystals were then cooled to 77 K, slowly to avoid cracking. The crystals were all irradiated in a ^{60}Co γ source to a total dose of 2–3 Mrad at 77 K. Care was taken so that no light reached the sample during and after irradiation, unless the sample was deliberately bleached. The ESR instrument was a Varian E-9 operated at X band. In order to observe anisotropic couplings the crystals were mainly rotated about an axis along the sample tube. However, some new techniques were also tested. It was possible to break off half a centimeter of the crystal and then to mount it in three mutually perpendicular orientations in the large cavity (Varian E-235) by using a simple holder and a slightly modified variable temperature device. The samples were bleached with filtered light from a tungsten lamp directly in the ESR cavity.

Results and Discussion

Methyl Acrylate

Most of the results on MA single crystals have been published earlier (Gillbro, Kinell & Lund, 1971), so only the most interesting findings will be noted here. Two different kinds of radical pairs were observed, both consisting of radicals formed from the monomer by hydrogen atom addition. The distance, r, between the radicals in one of the pairs was estimated from eq. (3) to be $r \leqslant 5.9 \pm 0.1$ Å. This distance corresponds to three possible intermolecular spacings between nearest-neighbour monomer molecules, according to the crystal structure (Brown, Gillbro & Nilsson, 1971). The lack of accuracy was due to the difficulty involved in rotating the crystal about three mutually perpendicular axes in the ESR cavity. In the present work the spectra were measured while the crystal was rotated about three mutually orthogonal directions. However, the maximum coupling was still observed when the crystal was rotated about the tube axis, and $d_{max} = 289$ G (Fig. 1) was found to correspond to $r \leqslant 5.8$ Å.

According to the crystal structure there is only one distance between closest neighbours which is shorter than 5.8 Å, and that distance is 4.55 Å. Assuming that crystal growth takes place along the crystallographic b axis, then this means that in the present case the tube axis is equivalent to the b axis, and the smallest angle between the magnetic field B and the vector r joining the paired radicals will be 26°10′ on rotation about the b axis. Using eq. (3) we obtain $d = 285$ G, which is in good agreement with experimental results. In

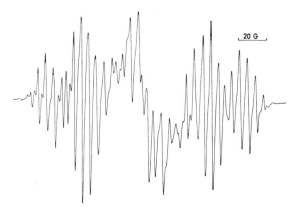

Fig. 1. ESR spectrum from a single crystal of methyl acrylate irradiated and recorded at 77 K in the dark. Dose 2.4 Mrad.

the crystal there are two symmetry-related pairs, which are magnetically nonequivalent, with the same separation, i.e. 4.55 Å.

In the previous work on MA no precautions were made to keep the sample in the dark following γ irradiation. In spite of this, the total number of radical pairs was estimated to be about 10% of the total number of unpaired monomer radicals in the sample. This time, however, extreme care was taken not to illuminate the irradiated MA crystal before recording the spectra. As a result the total number of radical pairs increased very much. As can be seen in Fig. 1, the number of radical pairs is in fact slightly greater than the number of unpaired radicals. When the sample was illuminated with unfiltered light for 30 s the amount of radical pairs decreased to 23.5% of its initial value, but no change in the central part of the spectrum was observed; this indicates that no new radicals were produced during the bleaching process. The two spectroscopically different types of radical pairs decreased in exactly the same way on bleaching. This means that they are chemically equivalent and that their different spectra are caused by their different orientations in the crystal lattice. This possibility was also suggested in the previous article from symmetry considerations.

Radical pairs are almost certainly produced in the heavily ionized tracks of the ionizing radiation. This conclusion is based on the experimental result that the percentage of radical pairs relative to the total number of radicals is independent of the radiation dose (Gillbro, Kinell & Lund, 1971). If the radical pairs were formed by diffusion of hydrogen atoms out of the tracks of the ionizing radiation and subsequent reaction with monomer molecules, the radical concentration would have a fairly homogeneous distribution throughout the crystal. The probability of radical pair formation in such a process is expected to be low at moderate doses and to increase with prolonged

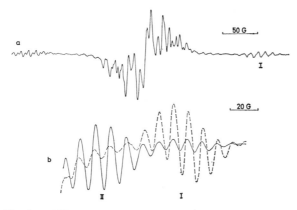

Fig. 2. *a*, ESR spectrum from a single crystal of methacrylonitrile irradiated and recorded at 77 K in the dark. Dose 3 Mrad; *b*, Dotted line: the high field (right) wing of Fig. 2 *a* at higher gain. Solid line: the same spectrum after illumination with filtered light ($\lambda > 4\,750$ Å) for 25 min.

irradiation. Although the mechanism for the formation of radical pairs is still not well understood, the following scheme (I) seems to hold for saturated hydrocarbons (Iwasaki, Ichikawa & Ohnori, 1969; Gillbro, Kinell & Lund, 1969).

$$\begin{vmatrix} RH \\ RH \end{vmatrix} \to \begin{vmatrix} RH^+ \\ RH^- \end{vmatrix} \to \begin{vmatrix} R^{\cdot} \\ \{RH_2^{\cdot}\} \end{vmatrix} \to \begin{vmatrix} R^{\cdot} \\ R^{\cdot} \end{vmatrix} + H_2 \tag{I}$$

It is tempting to suggest a similar mechanism for vinyl monomers, especially since methyl methacrylate and methyl acrylate are very strong electron scavengers and easily form anion radicals, which are quite stable in the glassy state (Gillbro, Yamaoka & Okamura, to be published). A proton transfer from the cation to the anion could be a possible explanation for the formation of hydrogen-addition type radicals. Since a lot of ions are formed close together in a spur, the probability of pair formation should be high.

MA is known to polymerize poorly in the solid state. This fact is of course mainly due to the crystal structure, which hardly permits any polymerization (Brown, Gillbro & Nilsson, 1971). It is also obvious that the considerable production of radicals in close pairs, which recombine easily on bleaching, will seriously decrease the remaining number of radicals that can initiate polymerization. One would expect to find a large quantity of MA dimers, $CH_3COO\ CH(CH_3) \cdot CH(CH_3)\ COOCH_3$, in the samples irradiated in the solid state and then warmed to the melting point.

Methacrylonitrile

The radiation chemistry of methacrylonitrile (MAN) is relatively unknown and only a few ESR studies have been reported (Bensasson et al., 1963; Mori,

Tabata & Oshima, 1970). In the work by Mori et al. (1970) a partly single-crystalline sample was used and the formation of radical pairs was assumed, although no $\Delta M_s = 2$ transition was observed, nor any hyperfine structure of the pair. Fig. 2*a* shows the best resolved spectrum when a MAN single crystal was rotated about the tube axis in the dark. The complex central part was not analysed, but on the wings there are two strongly anisotropic groups of lines. The hyperfine coupling constant within each group was measured and found to be 10.8 ± 0.2 G and the maximum splitting between the outermost groups was 351 G, which corresponds to a distance $r < 5.4$ Å between the two radicals in the pair. When the crystal was illuminated with infrared light ($\lambda > 4\,750$ Å) the radical pair (I) was bleached slowly and a new pair (II) appeared closer to the central spectrum, as illustrated in Fig. 2*b*. The pink colour of the sample disappeared on bleaching.

From measurements of the decrease of pair I and the increase of pair II on bleaching it was found that the increase of pair II was faster than the decrease of pair I. The maximum splitting of pair II (207 G) corresponds to $r < 6.45$ Å. Further rotation of the bleached crystal showed that yet another pair, III, was present with a maximum splitting of 228 G, which gave a calculated separation $r < 6.25$ Å between the radicals. The hfs coupling constant was 11 G, i.e. the same as that for pairs I and II. After bleaching, the central spectrum was somewhat simplified and in certain directions a seven-line spectrum with an hfs coupling constant of 21.5 G was recorded. Rotation of the crystal about two axes perpendicular to each other and to the tube axis gave no further information about the maximum distances in the radical pairs. After bleaching the weak $\Delta M_s = 2$ transition at half field was recorded and the peak separation was 11 G. However, because of the low intensity, the number of lines in this spectrum could not be evaluated.

The overlapping between ESR spectra from different radical pairs or between the central spectrum and the radical pair spectra was serious in most directions, and consequently the dipole-dipole coupling tensor (**D**) could not be evaluated; further, it was not possible to count the exact number of lines in the radical pair spectra. However, the intensity distribution around the central line in the spectrum of radical pair I (Fig. 2) was $1.0 : 0.885 : 0.595 : 0.25$ —, which is close to the binomial intensity distribution for a 13-line spectrum: $1.0 : 0.853 : 0.536 : 0.238 : 0.071 : 0.013 : 0.001$.

For pairs II and III, six lines were clearly observed on the high-field side of the central line in the radical pair spectra as shown in Fig. 3. The intensity distribution was for pair II: $1.0 : 0.847 : 0.524 : 0.26 : 0.08 : 0.03 : <0.01$. Consequently the total number of lines should be 13, taking into consideration the lines in the supposedly symmetrical spectrum which are overlapped by the central spectrum from unpaired radicals.

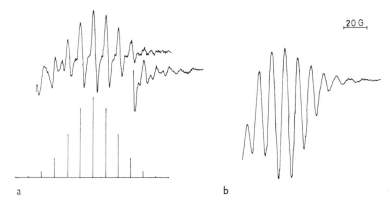

Fig. 3. *a*, The ESR spectrum from radical pair II in a bleached MAN single crystal. Measured at 77 K. The binomial distribution for a 13-line ESR spectrum is also shown for comparison; *b*, The ESR spectrum from radical pair III in a bleached MAN single crystal. Recorded at 77 K.

This number of lines together with the hfs coupling constant of 10.8 G, which is half of that found for the unpaired radical $CH_3\dot{C}(CH_3)CN$ (21.5 G), leads to the conclusion that the radical pairs I, II and III are all of the same type, i.e. that they consist of two $CH_3\dot{C}(CH_3)CN$ radicals. Only the distances, *r*, separating the monomer radicals are different.

Since new radical pairs are obviously produced during the bleaching process and since there is a simultaneous change in the central part of the MAN spectrum, there could be a connection between these two events. One possible explanation is that light induces a proton transfer between neighbouring MAN radical ions in a spur produced by the ionizing radiation. In this way several radicals of the hydrogen-addition type could be formed in close proximity to each other.

Conclusion

The production of radical pairs is of great importance in the radiation chemistry of vinyl monomers in the solid state. Thus, e.g., in methyl acrylate there are formed more paired than unpaired monomer radicals upon γ irradiation in the dark. The paired radicals are easily photobleached in methyl acrylate as well as in methacrylonitrile. In methacrylonitrile, however, new radical pairs are also produced on photobleaching. All radical pairs in methacrylonitrile obviously consist of two $CH_3\dot{C}(CH_3)CN$ radicals, only the distances separating the radicals in a pair are different.

Since many of the radicals in irradiated crystals of methyl acrylate, methyl methacrylate, acrolein and methacrylonitrile are trapped pairwise, one would expect a large quantity of dimers in the radiolysis products of these systems and the rate of polymerization would be low.

References

Adler, G. & Petropoulos, I. H., J. Phys. Chem., *69*, 3712 (1965).

Bamford, C. H., Eastmond, G. C. & Sakai, Y., Nature, *200*, 1284 (1963).

Bensasson, R., Bernas, A., Bodard, M. & Marx, R., J. Chim. Phys., *60*, 950 (1963).

Brown, A., Gillbro, T. & Nilsson, B., J. Polymer Sci. A-2, *9*, 1507 (1971).

Dahlgren, T., Gillbro, T., Nilsson, G. & Lund, A., J. Phys. Sci. Instr., *4*, 61 (1971).

Gillbro, T., Kinell, P. O. & Lund, A., J. Phys. Chem., *73*, 4167 (1969).

Gillbro, T., Kinell, P. O. & Lund, A., J. Polymer Sci., A-2, *9*, 1495 (1971).

Gillbro, T., Yamaoka, H. & Okamura, S., to be published.

Iwasaki, M., Minakata, K. & Toriyama, K., J. Chem. Phys., *54*, 3225 (1971).

Iwasaki, M., Ichikawa, T. & Ohmori, T., J. Chem. Phys., *50*, 1984 (1969).

Kurita, Y., J. Chem. Phys., *41*, 3926 (1964).

Lebedev, Ya. A., Radiation effects (ed. L. T. Chadderton) *1*, 213 (1969).

Mori, K., Tabata, Y. & Oshima, K., Kogyo Kagaku Zasshi, *73*, 2475 (1970).

O'Donnell, J. H., McGarvey, B. & Morawetz, H., J. Amer. Chem. Soc., *86*, 2322 (1964).

Shiga, T., Lund, A. & Kinell, P. O., Int. J. Radiat. Phys. Chem., *3*, 145 (1971).

Shioji, Y., Ohnishi, S. I. & Nitta, I., J. Polym. Sci., A, *1*, 3373 (1963).

Ueda, H., J. Polym. Sci., A, *2*, 2207 (1964).

Discussion

Henriksen

One important aspect of radical pair formation is whether the two individual species forming the pair are equivalent. Do you think there is a possibility for two different radicals in your experiments?

Gillbro

We have considered the possibility of different radicals in the pair, but we obtained the best agreement with the spectra when a pair consisting of two radicals with hydrogen added to the double bond was considered.

Chapiro

I would like to comment on three points.

(1) You assume that radical pairs are formed in the spurs produced by the electron. This should not give rise, however, to a high enough concentration of such pairs to account for your result in which the concentration of radical pairs is higher than that of random radicals.

(2) Dimers may arise in irradiated crystals by direct interaction of a triplet state with a neighbour molecule in the crystal (G.R.S. Schmidt and co-workers). Therefore the presence of dimers in the products of radiolysis would not be an unambiguous proof for the presence of radical pairs.

(3) The formation of radicals by H addition to the double bond raises a problem because the corresponding H-deficient radical has never been detected. Dr Marx has shown in the case of chloroacrylonitrile that the vinyl radical is stable (Marx, R, & Fenistein, S., J. Chim. Phys., *64*, 1424 (1967)) in the crystal. A mechanism of formation of the H˙ adduct radical was proposed involving a sequence of ionic processes.

Gillbro

(1) We think that the relative yield of radical pairs one could expect depends on the mechanism for energy absorption in the crystal and how this energy is distributed among the molecules in the spur. Since very little is known about these processes, especially in systems like organic crystals, we can only guess what might happen in the vinyl monomers studied.

One attractive feature in common for these molecules is that they contain a conjugated system, $C{=}C{-}C{=}O$ or $C{=}C{-}C{\equiv}N$. It is plausible to assume that energy absorbed by one molecule in such a conjugated system is easily distributed among several neighbouring molecules, which then form cations and anions situated close to each other in a spur. In our system this spur can consist of four molecules, which give rise to one radical pair. With this kind of mechanism the relative yield of random radicals does not necessarily have to be large.

(2) I agree that the presence of dimers in the products of radiolysis is not an unambiguous proof for the presence of radical pairs, but the presence of a high number of dimers of the kind, $CH_3CH(COOCH_3)CH(COOCH_3)CH_3$, as discussed in the paper, would strongly suggest that radical pairs might have recombined during melting of the MA crystal.

Lindberg

Your spectra are very well resolved. I think it would be possible by known theories to calculate the interaction energy between the parts in the radical pair and thus get an estimate regarding the distances. They can then be compared with X-ray data.

Gillbro

We have made that kind of calculations based on the dipole-dipole coupling between the radicals in a pair. However, since the full coupling tensor could not be evaluated in our experiment only a maximum distance separating the radicals could be calculated.

Hedvig

Would you consider a possibility that you observe spectra of colour centers formed after irradiation? Colour centres should exhibit hyperfine splitting due to the hydrogen atoms in the neighbourhood and correspondingly would look like radicals. This would explain the photobleaching effect you observed.

Gillbro

In our case the radicals behave mainly as point dipoles and we do not believe that the unpaired electrons are delocalized over many hydrogen atoms each giving such a high hfs as 11 G.

Ivin

I think that it would be feasible to do a chemically induced dynamic nuclear polarization (CIDNP) experiment starting with an irradiated crystal containing trapped free radicals and radical pairs. One would only need to observe the NMR spectrum (of the products) as the crystal melts and releases the trapped radicals. Possibly the simultaneous presence of solid material would present technical difficulties. Several possible types of NMR spectrum, with emission and absorption lines, might be observed, in accordance with the rules of CIDNP.

Williams

(1) Regarding the mechanism of electron rupture followed by proton transfer, one might have expected the following reactions to occur in MMA

$$CH_2{=}\overset{\overset{\displaystyle CH_3}{|}}{C}{-}\overset{\overset{\displaystyle O}{\|}}{C}{-}OCH_3 + e^- \rightarrow CH_2{=}\overset{\overset{\displaystyle CH_3}{|}}{C}{-}\overset{\overset{\displaystyle O^-}{|}}{\underset{\cdot}{C}}{-}OCH_3$$

$$\Big\downarrow {+}\,H^+$$

$$CH_2{=}\overset{\overset{\displaystyle CH_3}{|}}{C}{-}\overset{\overset{\displaystyle OH}{|}}{\underset{\cdot}{C}}{-}OCH_3$$

As Professor Szwarc has pointed out, this is the enol form of the radical you have proposed

$$(CH_3)_2\dot{C}{-}\overset{\overset{\displaystyle O}{\|}}{C}{-}OCH_3$$

Perhaps both of these radicals may be present.

(2) I think it is much more likely that radical pairs are formed in spurs rather than tracks if the mechanism of formation were to involve the ionization of two molecules at a short distance apart.

(3) Do you have some idea of the mechanism involved in the photobleaching of these radical pairs?

Gillbro

We consider the recombination of the radicals in a pair to be the most probable mechanism involved in the photobleaching.

Nature of the Paramagnetic Intermediates Formed in Initiating Reactions of γ-Irradiated Butadiene in the Crystalline, Adsorbed and Glassy States

By Anders Lund, Tetsuo Shiga and Per-Olof Kinell[1]

The Swedish Research Councils' Laboratory, Studsvik, Nyköping, Sweden

Introduction

Paramagnetic intermediates play an important role in the initiation reactions of irradiated monomer systems (Williams, 1968). Identification of the species then becomes important to clarify primary processes in radiation-induced polymerization. In the solid state the intermediates have a lifetime sufficiently long to permit observation after completing the irradiation. For this purpose electron spin resonance (ESR) is a powerful tool which provides direct information about the molecular structure and sometimes also about the orientation of species.

The work reviewed in this paper (Shiga, 1971; Shiga, Lund & Kinell, 1971, 1972) was undertaken in an effort to elucidate the nature of intermediates formed during irradiation of 1,3-butadiene. Several types of matrix were employed, depending on the specific problem under consideration. Single crystals of the pure monomer were used to investigate the geometry of the radicals by observing the angular dependence of the spectra. Some information regarding the direction of propagation could thus be inferred. Silica gel used as an adsorbent serves to stabilize cationic intermediates and also provides a means of obtaining better resolved spectra, by comparison with those obtained in polycrystalline and glassy matrices. These latter types of matrix were employed to study the selective formation and initiating action of negative and positive solute ions in different types of solvent. One further advantage of using glassy matrices is that scavenger additives can be used to inhibit one or other type of reaction. In this manner the reaction mechanism can be elucidated.

Experimental Section

The method of obtaining single crystals of compounds having subambient melting points has been described previously (Dahlgren et al., 1971). The materials were purified by distillation and drying under reduced pressure as de-

[1] Present address: Department of Physical Chemistry, University of Umeå, S-901 87 Umeå, Sweden.

tailed in the original papers (Shiga, Kinell & Lund, 1971, 1972). The samples which were prepared under vacuum and contained in Suprasil tubes were γ-irradiated at 77 K mostly to a total dose of about 1 Mrad at a dose rate of (0.7 ± 0.1) Mrad/h. The subsequent analysis was performed in the temperature range 77–300 K with a Varian type E-9 instrument operating at a frequency of 9.2 GHz.

Results

Single Crystal Butadiene

The spectra obtained at 77 K were difficult to interpret due to low resolution, which however improved when the sample was initially warmed. Measurements at 143 K with the crystal contained in a Suprasil tube demonstrated a pronounced angular variation of the spectrum when the sample was rotated about the container axis. A detailed analysis of the coupling tensors was attempted but has not yet been completed. Rotation of a piece of the crystal about a second axis at right angles to the tube axis and about a third axis mutually perpendicular to the two first axes indicates, however, that the maximum splitting from two equivalent protons is $a_1 = 18.5$ G. A second coupling has a maximum of $a_2 = 5$ G and a third approximately isotropic splitting is $a_\beta = 18.5$ G, Fig. 1. This corroborates an assignment to an allyl type of radical

in which the C–H bonds are parallel and the four β protons interact equally.

1,3-Butadiene Adsorbed on Silica Gel

Three kinds of radicals attributable to the hydrocarbon were observed at 77 K to an extent which varied with the experimental conditions. With a butadiene concentration of 0.1 % adsorbed on a gel previously dried at 500 or 650°C, two components were trapped. One of the components which disappeared at 195 K seemed to consist of five lines separated by 11 G, as revealed by a subtraction procedure. This absorption was absent in the presence of triethylamine, (TEA), which is known to be a positive charge scavenger. Several lines having peak-peak widths of only 1.0 G were resolved in a sample containing 0.1 % 1,3-butadiene-1,1,4,4-d_4. The spectrum could be interpreted with the parameters $a_H = 3.16$ G (2H) and $a_D = 1.74$ G (4D), the latter value

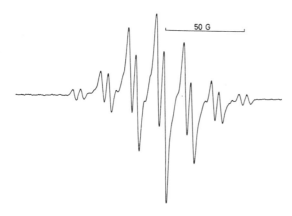

Fig. 1. Single crystal spectrum from irradiated 1,3-butadiene recorded at 143 K with the crystal oriented for a maximum separation of the lines.

corresponding to $a_H = 6.514 \times 1.74 = 11.3$ G. These couplings are in complete accordance with those estimated for the butadiene cation, employing the parameters previously derived for the negative ion in a molecular orbital calculation (Snyder & Amos, 1965).

After the cation had decayed the remaining signal changed reversibly with temperature in the range 77–200 K. When a gel which had been refluxed with heavy water was used as an adsorbent the sample containing 0.3 % butadiene gave a spectrum with six main components separated by the same splitting, 14 G, as in the seven line spectrum observed with the normal silica gel. This suggests that the hydrogen addition type radicals $CH_2D\dot{C}HCHCH_2$ and $CH_3\dot{C}HCHCH_2$ are formed on the deuterated and normal silica gel respectively. In an attempt to analyze the spectrum obtained at 193 K (Fig. 2) the following parameters were employed with a line shape simulation programme (Lefebvre & Maruani, 1965; Maruani, 1970): $a_1 = 13.8$ G, $a_2 = 3.8$ G, $A_{||} = 16.5$ G, $A_\perp = 13.0$ G (2H), $A_{||} = 17.5$ G, $A_\perp = 16.4$ G (3H).

At a higher butadiene content of 10 % a third type of spectrum was observed. Measurement at 195 K gave a well resolved spectrum which was attributed to the allyl type of radical $\dot{C}H_2$–CH=CH–CH$_2$–R following comparison with resonance data from the aqueous redox system containing 1,3-butadiene (Yoshida & Rånby, 1967). The reversible change with temperature indicates that this species is present immediately after irradiation at 77 K but in another conformation. A similar spectrum appeared when 2 % of butadiene was added to silica gel previously irradiated at 77 K and subsequently warmed to 195 K. This reaction occurred also when the silica gel sample was initially illuminated with filtered light ($\lambda > 515$ nm) in order to bleach one of the centres in the gel with $g = 2.0070$ (cf. Kinell et al. (1970)).

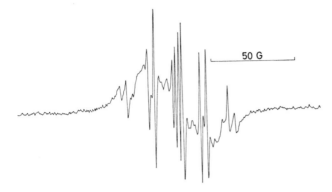

Fig. 2. Spectrum from the irradiated system 0.3 % butadiene/silica gel recorded at 193 K.

1,3-Butadiene in Organic Matrices

An absorption in the central part of the spectrum from the irradiated n-butyl chloride glass (BuCl) which remained when the solvent spectrum had decayed at 100 K is most probably derived from radiation-induced reactions with the solute. This absorption, which had a doublet splitting of 13.8 G when 1,3-butadiene-1,1,4,4-d_4 was used, is attributed to the allyl type radicals $\dot{C}H_2$–CH=CH–CH$_2$–R and $\dot{C}D_2$–CH=CH–CD$_2$–R respectively. The spectrum did not appear when 3.4 mole % TEA was present as a positive charge scavenger. This species is not formed in the polycrystalline carbon tetrachloride matrix; in this case the quintet structure with $a = 10.1$ G is most probably caused by the butadiene cation.

In the methyltetrahydrofuran (MTHF) glass the spectrum which remained after the solvent spectrum had been subtracted appeared to be formed in an anionic process. The evidence for this includes the inhibiting action when butyl chloride is present as an electron scavenger, the reduction of trapped electron signals in the presence of butadiene and the known ability of the MTHF matrix to favour anionic reactions (Tsuji et al., 1966). This species is also formed in the presence of TEA. An assignment based on reasonable estimates of anisotropic interactions in the butadiene anion trapped in a rigid matrix has been performed. Computer simulations employing the program MARU (Maruani, 1970) gave different powder spectra for the *trans* and the *cis* forms (Fig. 3b, c). The line shape (Fig. 3d) has the relative intensities $C_{cis} = 0.44$ $C_{trans} = 0.56$ as determined by a least squares fit to the spectrum of Fig. 3a (cf. Lund, 1972, for description of the method). Theoretical data (Snyder & Amos, 1965) provide rather weak evidence that the *trans* conformation is more stable than the *cis*.

A doublet structure similar to that observed in the butyl chloride matrix also appeared in glasses of 3-methylpentane and 3-methylhexane.

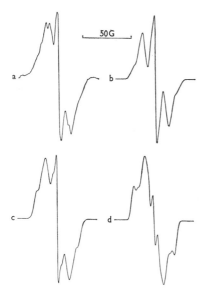

Fig. 3. *a*, Bleachable component in the spectrum from 3.0 mol % of 1,3-butadiene in MTHF containing 8.0 mol % TEA; *b*, *d*, Simulated powder spectra for the *trans* and *cis* 1,3-butadiene anion; *c*, Least squares fit of *a* to *b* (56 %) + *d* (44 %).

Discussion

The preceding results, summarized in Table 1, have demonstrated that depending on the conditions neutral radicals, anions and cations are formed from butadiene by γ-irradiation in the solid state. Since this diene might polymerize according to a free radical mechanism or in cationic or anionic reactions it is necessary to consider different alternatives for the initiating processes. A discussion based upon the nature of the matrices will be attempted below.

The fact that the spectra from the radical trapped in the pure hydrocarbon matrix at 143 K still feature anisotropy when the sample is rotated demon-

Table 1. *Paramagnetic Intermediates from Butadiene in Different Matrices*

Matrix	Initiating species	Propagating species	Remarks
Single crystal	Not identified	$-CH_2-\overset{\cdot}{C}H-CH=CH-CH_2-$	Cationic initiation assumed
Silica gel	$[CH_2CH\ CH\ CH_2]^+$	$\overset{\cdot}{C}H_2-CH=CH-CH_2-R$	
CCl$_4$	$CH_3-CH=CH-CH_3$, $[CH_2CH\ CH\ CH_2]^+$		Does not propagate Does not propagate
BuCl	Not identified	$\overset{\cdot}{C}H_2-CH=CH-CH_2-R$	Initiating species scavengable with TEA
MTHF	$[CH_2CH\ CH\ CH_2]^-$		Initiating species scavengable with BuCl
3 MP	Not identified	$\overset{\cdot}{C}H_2-CH=CH-CH_2-R$	Intensity reduced with TEA or BuCl
3 MH	Not identified		

strates that the radicals are located in an ordered way in the crystal structure. Thus, the polymerisation has not proceeded to any large extent since this would tend to destroy the crystalline order (Adler & Baysal, 1969). On the other hand the species observed cannot be formed from a single monomer unit, and it may therefore be concluded that the radical is located in oligomers. In the allyl type of radical having a *trans* conformation the maximum coupling of about 18 G occurs when the magnetic field is directed along the carbon chain axis (Heller & Cole, 1962). The observation of a maximum splitting of this magnitude in the propagating radical is consistent with a mechanism of formation through 1,4 addition of molecules located along the chain axis and possessing a *trans* geometry. Further support for this hypothesis can only be obtained when crystal structure data become available.

Tabata, Sobue & Oda (1965) have suggested that butadiene polymerizes cationically at 77 K. The following sequence of reactions is consistent with this view:

$$CH_2\!=\!CH\!-\!CH\!=\!CH_2 \rightsquigarrow [CH_2\!-\!CH\!=\!CH\!-\!CH_2]^+ + e^-$$

$$\xrightarrow{C_4H_6} CH_2\!=\!CH\!-\!\dot{C}H\!-\!CH_2\!-\!CH_2\!-\!CH_2\!=\!CH\!-\!CH_2^+$$

$$\xrightarrow{143\ K} CH_2\!=\!CH\!-\!CH_2\!-\!CH_2\!-\!\dot{C}H\!-\!CH\!=\!CH\!-\!CH_2^+$$

The isomerisation reaction is a suggested explanation of the irreversible change of spectra when the crystal is initially warmed.

Previous studies by Kinell et al. (1970) have shown that anions, cations and hydrogen-addition type radicals can be formed from molecules adsorbed on silica gel, very probably by reaction with electrons, positive charge and hydrogen atoms released from the irradiated adsorbent. The fact that the cation spectrum is observable at low butadiene content, but absent when the propagating radical occurs at high concentrations suggests that cationic initiation is the predominant process at 77 K. A similar process might take place on the preirradiated gel, resulting in the same type of propagating species. The following initiation reaction is consistent with observation.

$$S \rightsquigarrow S^\times$$

$$S^\times + CH_2\!=\!CH\!-\!CH\!=\!CH_2 \rightarrow S...[CH_2\!-\!CH\!=\!CH\!-\!CH_2]^{+}$$

$$S...[CH_2\!-\!CH\!=\!CH\!-\!CH_2]^{+} + CH_2\!=\!CH\!-\!CH\!=\!CH_2 \rightarrow$$

$$S...CH_2\!=\!CH\!-\!\dot{C}H\!-\!CH_2\!-\!CH_2\!-\!CH\!=\!CH\!-\!CH_2^+$$

Here $S^\times$ represents an active centre formed from the gel. The ESR spectra obtained following γ-irradiation at 77 K show that four main paramagnetic centres are formed (Kinell et al., 1970). Two of the centres were attributed to trapped electrons, one at $g=2.0008$ trapped at oxygen vacancy sites, the other at $g=2.0030$ trapped at a limited number of sites whose nature is not quite clear. Hole centres trapped at non-bridging oxygen atoms on the silica gel surface were thought to cause the absorption at $g=2.0070$, while the fourth type of centre at $g=2.011$ could be assigned to holes located close to ^{27}Al impurity atoms. The fact that the propagating radical was also formed on preirradiated silica gel after photobleaching apparently eliminates the $g=2.0070$ signal as the active centre. The trapped electron signal at $g=2.0008$ is not reactive towards adsorbed gases and can likewise be excluded. In the absence of experimental data for the reactivity of the other centres at $g=2.0030$ and $g=2.011$ the nature of the active sites can not be further elucidated.

The hydrogen-addition type of radical $CH_3-\dot{C}H-CH=CH_2$ is formed by the reaction with surface hydroxyl groups or adsorbed water molecules on the silica gel. The observation that it remains stable up to 200 K indicates that it does not easily take part in initiating reactions at low temperature. There is no experimental evidence for the presence of anionic species on silica gel or the initiation reactions of such species in the silica gel system.

The growing chains in a cationic mechanism may terminate by recombination with electrons. Then a biradical is formed which may end up in a final polymer molecule by hydrogen atom transfers along the chain. Crosslinking may be less probable in solid state systems.

The observation of the butadiene cation in the carbon tetrachloride matrix is in accordance with the expectation that alkyl halide matrices would favour cationic reactions. Thus, the propagating species observed in the butyl chloride glass is most likely formed in a cationic process. This is further supported by the inhibiting action of triethylamine which is known to be a positive charge scavenger.

The anions trapped in the MTHF glass do not initiate polymerisation. The propagating species observed in the alkane glasses might, however, form partly in an anionic process as evidenced by the partial scavenging by butyl chloride. Probably cationic initiation occurs simultaneously since TEA also reduces the intensity of the propagating species.

References

Adler, G. & Baysal, B., Mol. Cryst., *6*, 361 (1969).
Dahlgren, T., Gillbro, T., Nilsson, G. & Lund, A., J. Phys. E. Sci. Instr., *4*, 61 (1971).
Heller, C. & Cole, T., J. Chem. Phys., *37*, 243 (1962).
Kinell, P. O., Komatsu, T., Lund, A., Shiga, T. & Shimizu, A., Acta Chem. Scand., *24*, 3265 (1970).
Lund, A., J. Phys. Chem., *76*, 1411 (1972).

Maruani, J., extended and rewritten version (1970) of programme described by Lefebvre, R. & Maruani, J., J. Chem. Phys., *42*, 1480 (1965).

Shiga, T., Investigations on the initiating processes in radiation-induced ionic polymerization. Dissertation, Royal Institute of Technology, Stockholm (1971).

Shiga, T., Lund, A. & Kinell, P. O., Int. J. Radiat. Phys. Chem., *3*, 145 (1971).

Shiga, T., Lund, A. & Kinell, P. O., Int. J. Radiat. Phys. Chem., *3*, 131 (1971).

Shiga, T., Lund, A. & Kinell, P. O., Acta Chem. Scand., *25*, 1508 (1971).

Shiga, T., Lund, A. & Kinell, P. O., Acta Chem. Scand., *26*, 383 (1972).

Snyder, L. C. & Amos, T., J. Chem. Phys., *42*, 3670 (1965).

Tabata, Y., Sobue, H. & Oda, E., J. Phys. Chem., *65*, 1645 (1965).

Tsuji, K., Yamaoka, H., Hayashi, K., Kamiyama, H. & Yoshida, H., J. Polymer Sci., Part B, *4*, 629 (1966).

Williams, F., Principles of Radiation-Induced Polymerization in Fundamental Processes in Radiation Chemistry (ed. P. Ausloos). Interscience, New York 1968.

Yoshida, H. & Rånby, B., J. Polymer Sci., part C, 1333 (1967).

ESR Studies of Aliphatic Radical Anions and Radical-Anion Pairs in γ-Irradiated Crystalline Solids

By M. A. Bonin, Y. J. Chung, E. D. Sprague,[1]
K. Takeda,[2] J. T. Wang and F. Williams

Department of Chemistry, University of Tennessee, Knoxville, Tenn. 37916, USA

Introduction

It is well known that many organic molecules possessing low-lying orbitals form radical anions by electron attachment, and that several of these radical anions can initiate vinyl polymerization. However, there have been relatively few ESR studies of simple aliphatic radical anions, presumably because these species are generally too unstable to be observed under conventional experimental conditions in the liquid state. To some extent, this limitation set by the intrinsic chemical stability can be overcome by solid state studies at low temperatures, and this paper summarizes recent ESR work in this laboratory dealing with the identification and reactions of aliphatic radical anions in γ-irradiated crystalline solids. Incidental to this research, evidence has been obtained for weakly interacting radical-anion pairs formed by the dissociation of certain radical anions. Another aspect of this work which has received particular attention but will not be discussed here concerns the unexpected occurrence of hydrogen atom abstraction reactions by methyl radicals at 77 K (Sprague & Williams, 1971a). Detailed kinetic studies on a number of different systems (Wang & Williams, 1972) provide strong evidence for a large contribution from quantum mechanical tunneling in such reactions at low temperatures (LeRoy, Sprague & Williams, 1972).

Results and Discussion

Only a brief survey of the main findings will be given here, and many historical and experimental details are of necessity omitted. Since the results are highly characteristic for each compound and even for the particular crystalline state in which it is γ-irradiated, each system is discussed separately.

---

[1] Present address: Max-Planck-Institut für Kohlenforschung, Mülheim, West Germany.
[2] Present address: Fuji Photo Film Co. Ltd., Research Laboratories, Tokyo, Asaka, Saitama, Japan 351.

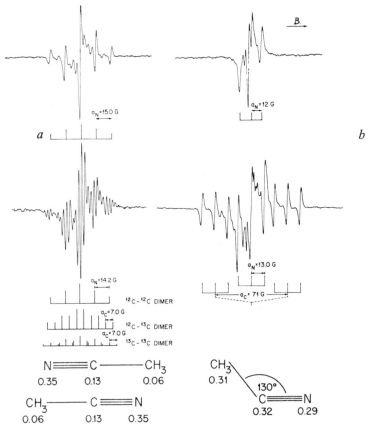

Fig. 1. ESR spectra and spin density distributions for dimer (*left*) and monomer (*right*) radical anions of acetonitrile. The upper and lower spectra were obtained using $CD_3C^{14}N$ and $CD_3{}^{13}C^{14}N$, respectively. The dimer and monomer radical anions were produced in Crystal I and Crystal II, respectively, of acetonitrile by γ irradiation at 77 K in the dark.

Acetonitrile Crystal I

The photobleachable colour center produced by γ irradiation of the meta-stabilized high-temperature phase of acetonitrile at 77 K has been identified as the dimer radical anion (Sprague, Takeda & Williams, 1971). All the ESR studies on the structure of the radical anions of acetonitrile have been carried out with deuterated compounds since the corresponding ESR spectra in CH_3CN are virtually unobservable because of proton hyperfine broadening and overlapping signals from other radicals. Representative ESR spectra from aligned crystals of $CD_3C^{14}N$ and $CD_3{}^{13}C^{14}N$ are presented in Fig. 1a and show hyperfine interaction of the unpaired electron with pairs of magnetically equivalent nuclei derived from two acetonitrile molecules. These results together with those for $CD_3C^{15}N$ and $^{13}CD_3C^{14}N$ establish that the dimer radical anion possesses a center of symmetry, as provided by an antiparallel placement of molecules.

The experimental spin densities derived from the principal values of the hyperfine tensors (Takeda, 1971) show that the unpaired electron resides mainly in a single 2p orbital on each nitrogen atom ($\Sigma\varrho_N = 0.68$) and the remaining fraction is accounted for by the spin densities on the methyl carbon ($\Sigma\varrho_C^{CD_3} = 0.13$) and on the cyanide carbon ($\Sigma\varrho_C^{CN} = 0.27$). Recent INDO calculations (Kerr, unpublished work) based on the structure shown in Fig. 1 are in good agreement with these results and confirm the previous assignment (Sprague, Takeda & Williams, 1971) of the nitrogen spin density to the in-plane 2p orbitals between the molecules.

Evidence which establishes the anionic nature of the species comes from the results of photobleaching and electron scavenging experiments. The purple colour and the ESR spectrum of the dimer species in CD_3CN are removed on bleaching with red or white light (Bonin, Tsuji & Williams, 1968), and the only product observable by ESR is the $CD_3^{\cdot}$ radical. Additional experiments with CH_3CN and $^{13}CD_3CN$ have verified that photodissociation gives the corresponding methyl radicals, so that the overall process can be represented by the equation,

$$(CD_3CN)_2^{\overline{\cdot}} \xrightarrow{\ h\nu\ } CD_3^{\cdot} + CN^- + CD_3CN$$

It is remarkable that this photobleaching reaction is thermally reversible, and the dimer radical anion recovers completely in CD_3CN. The reaction occurs slowly on standing at 87 K ($t_{1/2} \sim 12$ min) but takes place within a few minutes at 100 K ($t_{1/2} \sim 1$ min). On the other hand, the recovery of the dimer radical anion in CH_3CN is much less complete, particularly below 100 K, and some of the methyl radicals decay irreversibly. This difference between the reactions in CD_3CN and CH_3CN is due to the intervention of a competitive process of hydrogen atom abstraction by the methyl radical from a neighboring CH_3CN molecule (Takeda & Williams, 1970), a reaction which exhibits a very large primary deuterium isotope effect (Sprague & Williams, 1971). The kinetics of these reactions have been studied extensively (Bonin, 1969; Sprague, 1971; Wang, 1972).

Definite confirmation that the dimer species is an anion rather than a cation is supplied by the results obtained using CD_3CN doped with methyl halides as competitive electron scavengers (Sprague & Williams, 1971b). In mixtures containing 10 mol-% of CH_3Cl or CH_3Br, dimer radical anion formation is prevented and scavenging is complete as evidenced by the absence of colour and the lack of any photobleaching effect on the ESR spectrum of the sample after γ irradiation. Unexpectedly, the product of the scavenging reaction is of intrinsic interest. For the CH_3Br/CD_3CN mixture, the upper spectrum presented in Fig. 2a can be analyzed into two separate quartets of quartets as

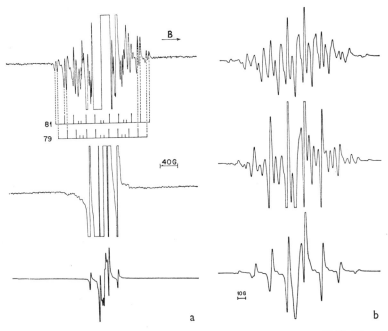

Fig. 2. *a*, ESR spectra of γ-irradiated 5 mole- % CH_3Br in CD_3CN at 88 K. The upper spectrum was recorded before annealing. The middle and lower spectra were recorded after the sample had been warmed to 175 K for a few seconds. For comparison, the upper and middle spectra were recorded at the same gain whereas the lower spectrum was recorded at half the modulation amplitude and one-tenth the gain. *b*, ESR spectra of γ-irradiated ca. 10 mol- % tert-butyl iodide in tert-butyl isocyanate at 77 K. The upper spectrum was recorded before annealing and the middle spectrum was recorded at the same gain setting after 1 min at 150 K. The lower spectrum was recorded at a reduced gain ($\times 1/3$) after prolonged annealing at ca. 150 K.

indicated by the stick diagram, and these can be assigned to CH_3 radicals interacting with bromide ions ($^{79}Br^-$ and $^{81}Br^-$ are present in almost equal abundance). When the sample was pulse annealed, the spectrum changed to that of the familiar quartet due to free CH_3 radicals (see Fig. 2*a*) showing that dissociation of the methyl radical-bromide ion pairs had occurred. As well as furnishing evidence that the paramagnetic dimer species in CD_3CN results from electron attachment, this experiment provides spectroscopic proof of dissociative electron capture by CH_3Br. Since the proton splitting in the spectrum of the unannealed sample is about 90 % of the value for the free CH_3 radical, it seems more appropriate to describe the original species as a methyl radical-bromide ion pair than a methyl bromide radical anion.

The incorporation of methyl isocyanide into acetonitrile Crystal I also resulted in efficient electron scavenging by the solute during γ irradiation, and at sufficiently high concentration of scavenger (10 mole % CH_3NC in CD_3CN), the dimer radical anion of acetonitrile was not produced. In this case CH_3 radicals were produced directly during γ irradiation so they must have originated from dissociative electron capture by CH_3NC (Wang, 1972). Although the reaction products are presumably identical to those (methyl radical and cyanide ion) produced on photodissociation of the acetonitrile dimer radical anion, no evidence was obtained for the generation of a photobleachable radical anion in a subsequent thermal recombination reaction. The loss of CH_3^- radicals by competitive hydrogen atom abstraction from CH_3CN (Wang & Williams, 1972) should be relatively unimportant in a largely deuterated matrix, so the failure to observe recombination can probably be attributed to the inverted position of the cyanide ion in the acetonitrile lattice.

Acetonitrile Crystal II

The lower crystalline phase of acetonitrile is prepared by slow cooling of the sample through a phase transition (Putnam, McEachern & Kilpatrick, 1965) at 215 K, about 12° below the melting point. On γ irradiation of this phase at 77 K, the monomer radical anion is produced (Takeda & Williams, 1969). The identification is based largely on ESR studies of γ irradiated $CD_3C^{14}N$ and $CD_3{}^{13}C^{14}N$. Although it is extremely difficult to grow single (aligned) crystals of this lower phase, this was achieved in some instances and particularly well resolved spectra were obtained in the case of $CD_3{}^{13}C^{14}N$.

A photobleachable triplet spectrum (Fig. 1b) is produced in $CD_3C^{14}N$ indicating that the unpaired electron interacts with only one ^{14}N nucleus. This is verified by the single crystal spectrum from $CD_3{}^{13}C^{14}N$ which is seen to consist of a doublet of triplets with a large ^{13}C splitting. The principal values of the $^{13}C(CN)$ hyperfine tensor in gauss are [71.7, 59.6, 53.0] from which the spin densities in the 2s and in-plane 2p orbitals of the cyanide carbon are calculated to be 0.054 and 0.19, respectively. These values give a ps hybridization ratio of 3.5 and by a well-known formula (Atkins & Symons, 1967) this leads to a calculated CCN angle of 130°. This result corresponds closely to the HCN angle of 131° in the isostructural molecule HCN^- (Adrian et al., 1969). The structure of the acetonitrile monomer radical anion is indicated in Fig. 1b and it should be noted that the total spin density on the carbon in the cyanide group ($\varrho_C^{CN} = 0.32$) is slightly larger than that on nitrogen $\varrho_N = 0.29$). This is expected in view of the electronegativity difference between carbon and nitrogen which should concentrate the charge distribution in the antibonding orbital more in favour of the carbon of the cyanide group. Another point of interest is the presence of substantial spin density on the carbon of the methyl group ($\varrho_C^{CD_3} = 0.31$). The large isotropic ^{13}C splitting of ~ 88 G from this carbon suggests that in the contributing structure $CD_3 \cdot CN^-$, the configuration at the methyl carbon atom is tetrahedral rather than planar.

The photobleaching and thermal recovery reactions of the acetonitrile monomer radical anion are strikingly similar to those already described for the dimer species in Crystal I, thereby providing strong confirmation of the radical anion assignment. Recovery of the monomer radical anion from the methyl radical produced on photodissociation proceeds to completion in CD_3CN as before, but the reaction rate for monomer recovery in Crystal II is about a factor of ten slower than that for dimer recovery in Crystal I at the same temperature although the activation energies have very nearly the same value (4.5 ± 0.5 kcal mol^{-1}) within experimental error. In the case of CH_3CN the recovery reaction competes with hydrogen atom abstraction by methyl radicals, again paralleling the results in Crystal I.

Adiponitrile

Monomer and dimer radical anions of $NC(CD_2)_4CN$ have also been identified by ESR. The monomer radical anion is formed by γ irradiation at 77 K of the crystalline phase produced directly from the melt whereas the dimer is produced on irradiation of the phase prepared by crystallization from the glass at low temperature. Excellent single crystals of the former phase were grown and enabled the determination of the principal values for the ^{14}N hyperfine tensor of the monomer radical anion (Takeda, 1971). These values in gauss are [21,0,0] which hardly differ from the corresponding results for HCN^- (Adrian et al., 1969). Only powder spectra could be obtained from the other crystalline phase but the parallel features of a quintet ESR spectrum are clearly displayed with $A_{||}$ (^{14}N) $= 17.2$ G which is almost identical to that found for the dimer radical anion in acetonitrile Crystal I. Both monomer and dimer radical anions are photobleached but only the monomer shows appreciable thermal recovery.

Succinonitrile

In contrast to the results for acetonitrile and adiponitrile, the dimer radical anion is formed in both crystalline phases of succinonitrile. Identification is based in each case on the powder ESR spectrum of succinonitrile-d_4 which closely resembles the corresponding spectra of the dimer radical anions of acetonitrile, propionitrile, and adiponitrile. The formation of the $\cdot CH_2CH_2CN$ radical is observed in both crystal phases of succinonitrile-h_4 after photobleaching but it is remarkable that thermal recovery to the dimer radical anion proceeds only in Crystal II (Campion & Williams, 1971). Similar results were obtained with succinonitrile-d_4. Presumably the radical produced on photobleaching in Crystal I relaxes to a position in the lattice which prevents the regeneration of the dimer radical anion.

Acrylonitrile

The ESR spectra of γ-irradiated acrylonitrile differ according to the nature of the crystalline phase (Chung, Takeda & Williams, 1970). A photobleachable

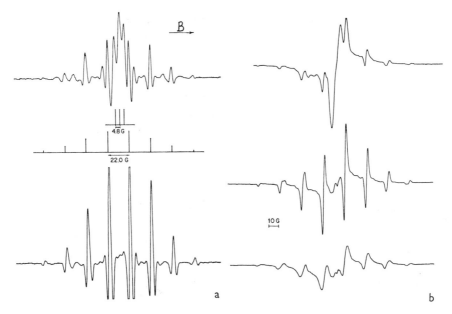

Fig. 3. *a*, ESR second-derivative spectra of γ-irradiated tert-butyl isocyanate at 77 K. The upper and lower spectra were recorded at the same gain settings before and after photobleaching with unfiltered tungsten light. *b*, ESR spectra of γ-irradiated crystalline tert-butyl isothiocyanate at 77 K. The upper spectrum was recorded during exposure of the sample to red light (Corning Filter No. 2030) and the middle spectrum during subsequent exposure to unfiltered tungsten light. The lower spectrum was recorded subsequently with the sample in the dark. These last two spectra were recorded at a reduced gain ($\times 0.8$).

triplet spectrum was observed in the low-temperature phase and has been tentatively assigned to the monomer radical anion. Only the spectrum of the CH₃ĊHCN radical could be identified in quenched samples of the high-temperature phase.

tert-Butyl Isocyanate

The ESR spectrum of crystalline tert-butyl isocyanate after γ irradiation (Fig. 3*a*) shows the well-resolved lines of the tert-butyl radical spectrum with a hfs of 22.0 G and three central components with a hfs of 4.8 G. The latter structure is selectively removed by photobleaching with visible light and this change is accompanied by a considerable increase in the intensity of the tert-butyl radical spectrum, as shown in the lower spectrum of Fig. 3A. By analogy to the studies on nitriles, the photobleachable species with the triplet spectrum can be assigned to the radical anion of tert-butyl isocyanate. Further evidence for this assignment comes from experiments using samples doped with tert-butyl iodide, an effective electron scavenger. As shown in Fig. 2*b*, a rather complicated ESR spectrum is obtained after such mixtures are γ irradiated at 77 K, and there is no photobleaching effect, suggesting that electron scavenging has occurred. On rapid annealing at higher temperatures, the spectrum gradually changes to that of the tert-butyl radical. Although the original spectrum of the doped sample is not easily analyzed, it can be interpreted reasonably well as the spectrum of tert-butyl radical-iodide ion pairs formed when tert-butyl iodide undergoes electron capture (Chung, 1972). The thermal dissociation of these radical-anion pairs parallels the results (Fig. 2a) for the methyl radical-bromide ion pairs in acetonitrile Crystal I.

tert-Butyl Isothiocyanate

As shown in the upper spectrum of Fig. 3b, the ESR spectrum of γ-irradiated tert-butyl isothiocyanate is composed of both a broad singlet feature and the multiplet spectrum of the tert-butyl radical. This particular spectrum was recorded while the sample was exposed to red light and the lines of the tert-butyl radical multiplet are much sharper than in the spectrum recorded with the sample in the dark, although the singlet feature was unaffected. This effect of line sharpening by red light was found to be completely reversible. On illuminating the sample with unfiltered tungsten light, the singlet spectrum was photobleached irreversibly and there was a large increase in the intensity of the tert-butyl radical spectrum, as shown in the middle spectrum of Fig. 3b recorded during this exposure to visible light. When the lamp was turned off, the lines of the tert-butyl radical spectrum broadened to give the lower spectrum. Again these lines could be sharpened by illumination with red light and the relative increase in signal height was similar to that observed before photobleaching. Further experiments showed that light in the near-infrared region ($\lambda > 1\ 250$ nm) was responsible for the photodynamic effect on the tert-butyl radical spectrum in this system (Chung & Williams, 1972).

On the basis of the photobleaching reaction, the singlet spectrum is assigned to the radical anion. Since the reversible line broadening is observed for the tert-butyl radicals produced by photodissociation of the radical anion as well as those formed directly by γ irradiation, this suggests that the broadening is due to the interaction of the tert-butyl radical with the thiocyanate anion. Moreover, a strong resemblance to the alkyl radical-halide ion pairs is indicated by comparison with the spectrum of the methyl radical-chloride ion pair which consists of the familiar quartet spectrum of the methyl radical broadened by hyperfine coupling with the chloride ion (Sprague, 1971). Although the exact nature of the magnetic resonance interaction responsible for line broadening in the tert-butyl radical-thiocyanate anion pair is not directly evident, the reversible photodynamic behavior can probably be attributed to motional effects which lead to an overall reduction in the g anisotropy (Chung & Williams, 1972).

Summary

This paper has been concerned with the identification and reactions of radical anions and radical-anion pairs derived by electron attachment to simple organic molecules. The chemistry observed in these systems can be conveniently summarized by the following set of basic processes, where R is an alkyl group and X is either a halogen (Cl, Br, I) or a pseudo-halogen (CN, NC, NCO, NCS). The analogous reactions involving dimer radical anions $(RX)_2^-$ can also be accommodated by this scheme.

Process		Example

Electron attachment $RX + e^- \longrightarrow$

$\nearrow R^\cdot + X^-$ (1 a) CH_3NC (in CD_3CN)

$\rightarrow [R^\cdot\text{----}X^-]$ (1 b) CH_3Br (in CD_3CN)

$\searrow RX^-$ (1 c) CH_3CN

Photodissociation $RX^- + h\nu$

$\nearrow R^\cdot + X^-$ (2 a) $(CH_3)_3CNCO$

$\searrow [R^\cdot\text{----}X^-]$ (2 b) $(CH_3)_3CNCS$

Thermal dissociation $[R^\cdot\text{----}X^-] \overset{\Delta}{\longrightarrow} R^\cdot + X^-$ (3 a) $(CH_3)_3CI$ (in $(CH_3)_3CNCO$)

Thermal recovery $R^\cdot + X^- \overset{\Delta}{\longrightarrow} RX^-$ (4) CH_3CN

A few generalizations can be made about the scope of these reactions. Electron attachment to alkyl halides is generally found to be dissociative giving rise to either separated fragments as in (1a) or to a radical-anion pair as in (1b). On the other hand, electron attachment to molecules possessing π electron systems is predominantly non-dissociative, as in (1c), although CH_3NC constitutes an exception to this trend. The interconversions of the different paramagnetic species as represented by reactions (2 a), (2 b), (3 a), and (4) are of particular interest. Both the radical anion RX^- and the radical-anion pair $[R\cdot\text{----}X^-]$ can be regarded as intermediate stages in the overall process of dissociative electron attachment. However, if the dissociation of the radical anion is an endothermic process, the fragments produced by photodissociation in (2a) can recombine thermally in (4) to regenerate the radical anion. This type of behavior is exemplified by the nitrile systems.

There is a certain arbitrariness about defining a radical-anion pair in crystalline solids according to the observation of a magnetic resonance interaction between the nominally paramagnetic and diamagnetic fragments. At present it is impossible to state whether the radical-anion pair represents a true minimum in the potential energy curve or merely reflects constraints imposed by the crystalline lattice.

This research was supported by the U. S. Atomic Energy Commission and this is document No. ORO-2968-76. Thanks are due to Dr C. M. L. Kerr for helpful discussions.

References

Adrian, F. J., Cochran, E. L., Bowers, V. A. & Weatherley, B. C., Phys. Rev., *177*, 129 (1969).

Atkins, P. W. & Symons, M. C. R., Structure of Inorganic Radicals, p. 257. Elsevier, Amsterdam 1967.

Bonin, M. A., Ph. D. Thesis, The University of Tennessee 1969.

Bonin, M. A., Tsuji, K. & Williams, F., Nature, *218*, 946 (1968).

Campion, A. & Williams, F., J. Chem. Phys., *54*, 4510 (1971).

Chung, Y. J., Ph. D. Thesis, The University of Tennessee, 1972.

Chung, Y. J. & Williams, F., J. Phys. Chem., *76*, 1792 (1972).

Chung, Y. J., Takeda, K. & Williams, F., Macromolecules, *3*, 264 (1970).

Le Roy, R. J., Sprague, E. D. & Williams, F., J. Phys. Chem., *76*, 546 (1972).

Putnam, W. E., McEachern, D. M., Jr. & Kilpatrick, J. E., J. Chem. Phys., *42*, 749 (1965).

Takeda, K., Thesis, Kyoto University 1971.

Takeda, K. & Williams, F., Mol. Phys., *17*, 677 (1969).

Takeda, K. & Williams, F., J. Phys. Chem., *74*, 4007 (1970).

Sprague, E. D., Ph. D. Thesis, The University of Tennessee 1971.

Sprague, E. D. & Williams, F., J. Amer. Chem. Soc., *93*, 787 (1971 *a*).

Sprague, E. D. & Williams, F., J. Chem. Phys., *54*, 5425 (1971 *b*).

Sprague, E. D., Takeda, K. & Williams, F., Chem. Phys. Lett., *10*, 299 (1971).

Wang, J. T., Ph. D. Thesis, The University of Tennessee 1972.

Wang, J. T. & Williams, F., J. Amer. Chem. Soc., *94*, 2930 (1972).

Discussion

Kinell

You mentioned that the dimer anion radical of CH_3CN had an antisymmetric structure with a distance between the central carbon atoms of 2.3 Å. Is this structure the energetically most stable one?

Williams

The structure mentioned in our paper is the one that gives calculated spin densities by the INDO method in closest agreement with the experimental results, and does not correspond to the lowest energy of the system. However, this structure is close to a position for which a minimum energy is found. In fact, several energy minima are found when the relative positions of the molecules are varied for a fixed geometry of the individual linear molecules, and two different molecular geometries give similar results. I should add that these calculations were carried out by Dr Carolyn M L Kerr and I am indebted to her for permission to quote these results.

Henriksen

I assume that your INDO calculations yield values for the isotropic hyperfine splitting of nitrogen. How is the correlation between the calculated values and those obtained by ESR? The reason I ask is that our experience with the INDO method so far seems to indicate that the calculated nitrogen splitting very often is in disagreement with that found by ESR.

Williams

The experimental results and the INDO calculations agree in showing that for both monomer and dimer radical anions, most of the total spin density on nitrogen is located in the 2p orbitals. Regarding the very small spin densities (0.01–0.02) in the nitrogen 2s orbital, the calculated isotropic hyperfine splitting is indeed found to be somewhat lower than the corresponding experimental result in each case. However, we do not view this particular comparison as a crucial test of our structural assignments.

Hedvig

What is the reason for the large isotope effect in hydrogen abstraction?

Williams

We have found that the kinetic isotope effect, k^H/k^D, for the abstraction reaction in acetonitrile Crystal I at 87 K is at least 2×10^3, a factor of 3 greater than the maximum isotope effect which can be calculated according to the difference in the initial zero point energies. This result together with other evidence (Le Roy, R. J., Sprague, E. D. & Williams, F., J. Phys. Chem, *76*, 546 (1972), Wang, J. T. & Williams, F., J. Amer. Chem. Soc., *94*, 2930 (1972)) strongly suggests that there is a large contribution from quantum mechanical tunneling in this H-atom abstraction reaction at low temperatures.

Sohma

I should like to make a comment concerning the narrowing of ESR spectrum caused by illumination of red light.

We observed a similar effect on the cyclopentanyl radicals. The ESR spectrum of this radical became narrower when it was illuminated by visible light at 77 K. The same narrowing effect was observed when the whole sample was warmed up to the elevated temperatures, say 130 K. The total energy of the light was not large enough to warm up the sample to such higher temperature. Thus, the limited region, where the radicals were trapped, was heated up by the selective absorption of the light through the same excitation of the radical. We call this phenomenon the local heating by the light. However, we do not know the molecular mechanism of this local heating.

Williams

The results mentioned by Professor Sohma are very interesting. Also in our system we cannot be sure of the photo-excitation mechanism responsible for the motional narrowing. We find that the relaxation time for line broadening in the dark is of the order of minutes which seems to be unusually long for a simple molecular excitation mechanism.

POLYMERIZATION REACTIONS

Radiation-Induced Polymerizations in the Vicinity of Glass Transition Temperatures

By Adolphe Chapiro

Laboratoire de Chimie des Radiations du CNRS, 92-Bellevue, France

In the past years several studies of this laboratory were devoted to the radiation-induced polymerization of monomer-solvent mixtures which form glasses at low temperatures. Some of the most significant results obtained in this work are reviewed in this communication with special emphasis on the behaviour and the nature of the active species trapped in irradiated glasses.

Vitreous Mixtures of Monomers with Mineral Oil

Earlier work, conducted with monomers dissolved in mineral oil and other oily compounds revealed that a very fast polymerization occurred below the glass transition temperature of the system. This effect was observed with many monomers, e.g. methyl methacrylate (Amagi & Chapiro, 1962; Chapiro & Pertessis, 1964; Chapiro & Roussel, 1967), methyl acrylate (Chapiro & Pertessis, 1964), vinyl chloride (Chapiro & Roussel, 1967), styrene (Chapiro & Pertessis, 1964; Chapiro & Roussel, 1967), acrylonitrile (Chapiro & Roussel, 1967), lauryl methacrylate (Chapiro & Perec, 1966). It was found, however, that the reaction which was fastest in highly diluted mixtures was actually a liquid phase polymerization taking place in small monomer droplets which separate from the oily solution on cooling and remain in a super-cooled state at very low temperatures (Spritzer, Sella & Chapiro, 1965, 1966; Chapiro & Nakashio, 1966).

The size of the droplets decreased with the monomer content in the mixture (Figs. 1, 2) and the rate of polymerization was found to be maximum for 10 % monomer solutions when the size of the droplets was 60 to 100 Å. A comparison of the observed rate at $-78°C$ with the value obtained by extrapolating bulk polymerization rate data to this low temperature showed that the reaction in isolated droplets in mineral oil proceeded ca. 1 000 times faster than in bulk at the same temperature (Chapiro & Roussel, 1967; Chapiro & Nakashio, 1966). This result was interpreted by analogy with emulsion polymerization by assuming that at a given time each droplet only contained one growing chain and that termination was controlled by the slow diffusion of free radicals

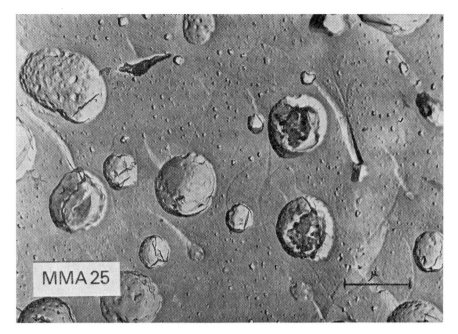

Fig. 1. Electron micrograph of a vitreous mixture of 25 % MMA with mineral oil fractured at −196°C (Spritzer, Sella & Chapiro, 1966).

formed in the mineral oil phase towards the droplets. ESR measurements (Marx, 1965) showed that upon irradiation, a 10/90, methyl methacrylate/ mineral oil mixture at −196°C essentially leads to alkane radicals. However, when this mixture is warmed to −130°C, the signal of alkane radicals suddenly changes to that of growing methyl methacrylate chains, indicating that diffusion of the radicals in the vitreous mineral oil actually takes place towards the monomer droplets. This diffusion is, however, a slow process and a chain initiated in a droplet may grow for a considerable length of time before it is terminated by a new diffusing-in radical. This slow termination accounts for the very high rate observed under such conditions.

Glasses Formed in Binary Mixtures of Crystallisable Compounds

More recent work was directed to the investigation of binary mixtures involving a crystallisable monomer and a crystallisable solvent which form, however, homogeneous glasses over a certain range of compositions. In such systems the polymerization was found to proceed either via free radicals or via ionic mechanisms depending on the nature of the system. The results are summarized in Table 1.

Experiments conducted at different temperatures conclusively demonstrated

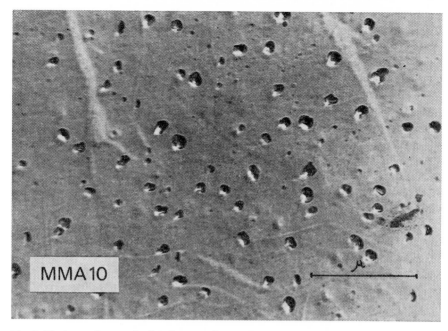

Fig. 2. Electron micrograph of a vitreous mixture of 10 % MMA with mineral oil fractured at −196°C (Spritzer, Sella & Chapiro, 1966).

that no significant chain propagation occurs in the vitreous state but that very favorable conditions for polymerization are met in a narrow range of temperatures above the glass transition temperature (T_g) of the mixture. The rate was found to exhibit a sharp maximum at a temperature T_{max} lying 5 to 10°C above T_g (see Table 1). A typical example of this behaviour is shown in Fig. 3

Table 1. *Mechanisms of Radiation-induced Polymerization in the Vicinity of T_g for various Systems*

Binary mixture	Monomer concentration %	T_g °C	T_{max} °C	Propagation mechanism	Reference
Acrylonitrile–DMF	30	−156	−150	Anionic	Perec, 1969
Acrylonitrile–isopropylamine	40	−160	−155	Anionic	Perec, 1969
Acrylonitrile–triethylamine	15	−157	−150	Anionic	Perec, 1969
Acrylonitrile–aniline	40	−130	−110	Free radical	Azikonda & Perec, 1969
MMA–aniline	50	−115	−95	Free radical	Azikonda & Perec, 1969
MMA–DMF	50	−140	−130	Free radical	Azikonda & Chapiro, 1971
Styrene–aniline	30	−98	—	Free radical	Azikonda & Perec, 1969
Styrene–methylene chloride	20 to 40	−60	—	Cationic	Perec, 1969
Styrene–hexachlorobutadiene	35	−120	−120	Cationic	Perec, 1969

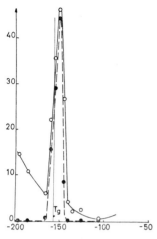

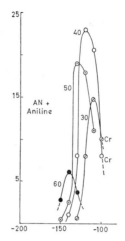

Fig. 3   Fig. 4

Fig. 3. *Abscissa:* temperature (°C); *ordinate:* % conversion (0.5 Mrad).
Influence of irradiation temperature on the rate of polymerization of a 30 % acrylonitrile solution in DMF. ○—○, irradiated mixtures warmed to room temperature under vacuum; ●--●, irradiated mixtures warmed in the presence of acetone (Perec, 1969).

Fig. 4. *Abscissa:* temperature (°C); *ordinate:* % conversion (6 h).
AN %/T_g°C: 60/–149; 50/–140; 40/–130; 30/–120.

which represents the rate of polymerization of a 30 % acrylonitrile solution in dimethylformamide as a function of irradiation temperature. A very similar behaviour is found with most other polymerizing mixtures, whether the reaction proceeds via ions or via free radicals. Fig. 4 represents the data obtained with acrylonitrile-aniline mixtures of different compositions. In each case the maximum rate occurs a few degrees above T_g.

The sudden rise in rate above T_g is undoubtedly due to a gradual increase of the mobility of monomer molecules as the system changes from a glass to a viscous liquid. A further increase in the temperature results in a sudden drop in the polymerization rate which is caused in certain systems by the further reduction of viscosity and the onset of bimolecular chain termination. In other mixtures (methyl methacrylate + dimethylformamide) (Azikonda & Chapiro, 1971) crystallization takes place at this point and this effect is also responsible for the sudden reduction in polymerization rate.

No significant chain propagation is detectable below T_g. The portion of the solid curve in Fig. 1 in the range of temperatures of -196 to -160°C corresponds to polymer obtained when the irradiated samples are warmed to room temperature under vacuum. This polymer is however formed *after irradiation*, during the warming period, since no polymer arises if acetone (an ionic inhibitor) is added to the irradiated mixture before warming the same (broken curve in Fig. 3) (see also p. 121).

Post-Polymerization in the Vicinity of T_g

Free Radical Processes

In free radical polymerization, growing chains are known to exhibit prolonged lifetimes when the viscosity of the reaction medium rises and thereby reduces the rate of chain termination. This effect, which occurs when polymer accumulates in the reaction mixture is usually designated as the "gel effect". ESR studies (Atheron, Melville & Whiffen, 1958) have confirmed the low rate of chain termination under such conditions. A similar effect arises in the vicinity of T_g where the mixture also exhibits a very high viscosity. Moreover, if the mixture is irradiated in the vitreous state, below T_g the resulting radicals remain trapped and can initiate a post-polymerization when the samples are warmed to room temperature (Chapiro & Nakashio, 1966). This post-polymerization upon warming may be prevented however in mixtures which tend to crystallize when warmed above T_g (Azikonda & Chapiro, 1971).

Ionic Post-Polymerization

Very little information is available on the lifetime of ionic growing chains trapped in glasses. A detailed study was therefore undertaken on the post-polymerization of acrylonitrile in its vitreous mixtures with dimethylformamide and triethylamine (TEA) (Bhardwaj, Chapiro & Perec, 1971). Mixtures of 30 % acrylonitrile with 70 % dimethylformamide and 15 % acrylonitrile with 85 % triethylamine were irradiated to a constant dose of γ rays at $-196°C$ and thereafter stored in a cryostat at a fixed temperature. After various storage times the samples were warmed to room temperature under standard conditions and the polymer was separated.

Results obtained with acrylonitrile–dimethylformamide mixtures at different temperatures are summarized in Fig. 5. It can be seen that after storage at -196 or at $-170°C$ the amount of polymer formed on warming on behalf of the trapped ionic species gradually decreases. As expected, the decay of trapped charges occurs faster at -170 than at $-196°C$. (The glass transition temperature is $-156°C$ for this mixture).

The curve obtained at $-150°C$ exhibits an odd shape, with a pronounced maximum obtained after ca. 5 h storage. At this temperature chain propagation sets in and the curve shown in Fig. 5 is a superimposition of the post-polymerization occurring at $-150°C$ and of the additional reaction occurring on warming. This second process can be prevented by warming the mixture in the presence of acetone, an inhibitor of the ionic chain propagation. Curve *a* in Fig. 6 shows the post-polymerization obtained under these conditions. After a short induction period, corresponding to the temperature rise of the

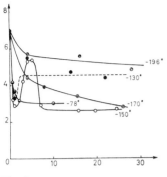
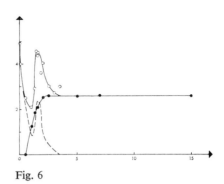

Fig. 5   Fig. 6

Figs 5, 6. *Abscissa:* time (h); *ordinate:* % conversion.

Fig. 5. Influence of storage time at different temperature on the amount of polymer formed after warming to room temperature a 30 % acrylonitrile solution in DMF irradiated to 0.2 Mrad at −196°C (Bhardwaj, Chapiro & Perec, 1971).

Fig. 6. Influence of storage time at −150°C on the amount of polymer formed after warming to room temperature a 30 % acrylonitrile solution in DMF irradiated to 0.09 Mrad at −196°C (Bhardwaj, Chapiro & Perec, 1971).

Curve a (●—●), mixture warmed in the presence of acetone;
Curve b (○—○) mixture warmed under vacuum.
Curve c (----) difference between ordinates of curves *a* and *b*.

mixture from −196°C to −150°C the post-polymerization sets in and proceeds for ca. 3 h. The additional post-polymerization on warming is shown by the broken curve, which is obtained by subtracting the ordinates of curve *a* from those of curve *b*. It appears that the amount of polymer formed on warming is not constant but rises with the extent of post-polymerization at −150°C. This undoubtedly follows from the increased viscosity of the reaction mixture due to polymer accumulation and this leads to slower chain termination and therefore to longer kinetic chains on warming.

Formation of "Complexes"

The fact that mixtures of two crystallisable compounds do not crystallise but form glasses over a certain range of concentrations indicates the formation of plurimolecular aggregates which are responsible for a rise in the viscosity of the mixture. This effect may lead to unusual reaction products as will be shown below.

The type of association involved is not firmly established. Depending on the system, one may consider dipole–dipole interaction, H-bonding or acid-base type complexes. Direct evidence of such aggregates is provided by viscosity measurements. Thus, acrylonitrile-dimethylformamide mixtures exhibit an almost linear relationship between viscosity at 25°C and composition; but

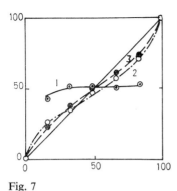 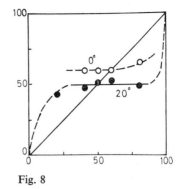

Fig. 7   Fig. 8

Figs. 7, 8. Mol-% MMA, *abscissa:* in monomers; *ordinate:* in copolymer.

Fig. 7. MMA content in copolymer as a function of styrene-MMA feed.
Curve 1 (○—○) 50% monomer solution in aniline irradiated at −95°C; *Curve* 2 (○-··-○) 50% monomer solution in aniline irradiated at +20°C; *curve* 3 (●--●), "normal" free radical copolymer (Azikonda & Chapiro, 1971, 1972).

Fig. 8. MMA content in copolymer as a function of styrene-MMA feed obtained by irradiating 50% monomer solutions in DMF at 0°C (○--○) and 20°C (●—●) (Azikonda & Chapiro, 1972).

at −8°C a pronounced maximum appears in the viscosity for acrylonitrile contents of 10 to 90% (Spritzer, 1971) and on further cooling, the solutions which contain 25 to 60% acrylonitrile do not crystallize but form glasses (Perec, 1969). This suggests the existence of plurimolecular aggregates presumably resulting from dipole–dipole interaction.

The polymerization mechanisms established for the various systems which are summarized in Table 1 are based on the influence of selective inhibitors such as particularly water or acetone for ionic polymerizations. Copolymerization studies conducted in these systems showed, however, a more complex behaviour. Thus, styrene and methyl methacrylate both polymerize via free radicals in aniline glasses. However, the composition of the copolymer formed in the same solvent in the vicinity of T_g remains almost constant as a function of monomer feed and is close to equimolecular with respect to both comonomers (Azikonda & Chapiro, 1971; 1972). This indicates the formation of an alternating copolymer. These results are summarized in Fig. 7. The same monomer-pair in dimethylformamide leads to an even more surprising behaviour (Azikonda & Chapiro, 1971; 1972). Both monomers polymerize in dimethylformamide mixtures in the vicinity of T_g if present alone. If, however, a few percent of styrene are added to methyl methacrylate or a few percent methyl methacrylate to styrene a complete inhibition occurs. At higher temperatures, copolymerization takes place but the resulting copolymer again exhibits a constant composition, the methyl methacrylate content being approx. 60% at 0°C and 50% at 20°C (see Fig. 8). This constant composition cannot result from a normal copolymerization process in which both monomers compete in their reaction with the growing chains. It thus appears likely that the regularly ordered sequence of monomer units in the copolymer is predetermined to a certain degree by some regular structure of the molecular aggregates. The fact that such ordered structure is retained at room temperature is surprising. Alternating copolymers were also found to arise in mixtures of acrylic acid with N-vinylcarbazole which form glasses at low temperature (Mankowski, unpublished). In this system the molecular aggregates presumably involve acid-base type complexes.

Conclusion

The various studies conducted on the polymerization of monomers in glass-forming media lead to a number of conclusions of general character:

1. Irradiation of glasses results in the formation of radicals and ionic centers which remain trapped and only slowly recombine at temperatures below T_g.

2. In certain mixtures, phase separation occurs on cooling and the monomer may then become dispersed in small droplets which remain in a supercooled liquid state down to very low temperatures. In such systems the polymerization is accelerated because chain termination is then diffusion controlled in a very viscous medium.

3. In homogeneous glasses, no significant chain propagation occurs in the vitreous state in view of the low mobility of monomer molecules.

4. Very favorable conditions for chain polymerization are met a few degrees above T_g when propagation becomes possible while termination is still hindered by the high viscosity of the medium.

5. Post-polymerization is observed in many mixtures on warming above T_g after irradiation, provided the system does not crystallize.

6. Glass forming mixtures composed of two crystallizable compounds often contain plurimolecular aggregates of more or less ordered structure. Alternating copolymers are found to arise in such mixtures.

References

Amagi, Y. & Chapiro, A., J. Chim. Phys., *59*, 537 (1962).
Chapiro, A. & Pertessis, M., J. Chim. Phys., *61*, 991 (1964).
Chapiro, A. & Roussel, D., J. Polymer Sci. Part C, *16*, 3011 (1967).
Chapiro, A. & Perec, L., J. Chim. Phys., *63*, 842 (1966).
Spritzer, C., Sella, C. & Chapiro, A., Compt. Rend. *260*, 2789 (1965).
Spritzer, C., Sella, C. & Chapiro, A., Compt. Rend., *261*, 1275 (1966).
Chapiro, A. & Nakashio, S., J. Chim. Phys., *63*, 1031 (1966).
Marx, R., J. Chim. Phys., *62*, 767 (1965).

Perec, L., J. Chim. Phys., *66*, 1742 (1969).

Azikonda, L. & Perec, L., Kinetics and Mechanisms of Polyreactions, vol. 4, p. 86. IUPAC, Budapest 1969.

Azikonda, L. & Chapiro, A., Third Symposium on Radiation Chemistry, Tihany, Hungary 1971. Preprint B/1.

Atherton, N. M., Melville, H. W. & Whiffen, D. H., Trans. Faraday Soc., *54*, 1300 (1958).

Bhardwaj, I. S., Chapiro, A. & Perec, L., Europ. Polymer J., *7*, 135 (1971).

Bhardwaj, I. S., Chapiro, A. & Perec, L., 2nd International Symposium on Chemical Transformations of Polymers. Bratislava, 1971, Preprint P-45.

Spritzer, C., J. Chim. Phys., *68*, 340 (1971).

Azikonda, L. & Chapiro, A., International Symposium on Macromolecular Chemistry. IUPAC, Helsinki, July 1972.

Mankowski, Z., Unpublished work in this laboratory.

Discussion

Charlesby

The reduction in the amount of polymer formed when the sample, irradiated at $-190°C$, is kept for a time at this temperature, is ascribed to the gradual loss of some ionic species, even at this low temperature. A related observation we often find in thermoluminescence is the emission of light in irradiated organics, when these are kept at liquid nitrogen temperature (preglow). This is ascribed to the slow leakage of electrons from their traps, to positive centres to give off light. It would be interesting to see whether these two effects can be correlated quantitatively, e.g. by the time dependence of this decay.

The use of very small polymerisable volumes, dispensed in a relatively inert matrix, should give some idea of the distribution of ionisation events due to gamma irradiation. For example if there are several ionisations in very close proximity within each spur, many of these volumes will contain pairs of ionisations, which react rapidly with each other and are relatively ineffective to give polymer. If, however, ionisations occur much further apart (i.e. the spur is large) most polymerisable volumes contain zero or one ion only, leading to high molecular weight polymer. In polymerisation of methyl methacrylate, we have measured the molecular weight distribution by gel permeation chromatography. The distribution appears to consist of two exponentials of very different averages indicating the existence of two polymerisation mechanisms, though there is little evidence of any abnormally high fraction of very low molecular weight polymer, as might be expected if radicals or ions are not distributed at random but in close proximity.

Hedvig

If a glass is irradiated below the glass transition temperature and consequently heated up a thermostimulated current peak is observed below T_g. This roughly corresponds to the thermoluminescence peak mentioned by Professor Charlesby. The difference is, however, that all the detrapped charge carriers contribute to the thermostimulated peak and only a fraction of them do so to the thermoluminescence peak. It would be interesting to see if there is a correlation between the initiation of the polymerization and the detrapping of charge carriers.

Chapiro

The "decay curves" for polymer formation at -196–$170°C$ as shown in Fig. 5, should not be considered as representing accurately the decay of charged active species. These curves represent a trend which actually conforms with decay curves determined by other methods. However, the amount of polymer formed on warming depends to a certain extent on the degree of conversion. This is particularly clear for the experiments conducted at $-150°$ (Figs 5, 6). It follows that the number of trapped active species cannot be strictly derived from the post-polymerization data.

Sohma

How did you determine the temperature of crystallization of the small droplets in the glassy matrix?

Chapiro

The crystallization temperature of monomer droplets dispersed in the vitreous mineral oil was measured using a cryostat with programmed temperature and observing the sample with a microscope under crossed polarizers. The crystallization corresponds to a sudden illumination of the field of the microscope and the corresponding temperature can be determined within a few $°C$ (see also Spritzer, Sella & Chapiro (1966); Spritzer (1971)).

Tsuji

You got alternating copolymers when MMA and St in DMF was irradiated at low temperature, and it was attributed to the aggregation of two monomers by dipole–dipole interaction.

Do you expect that alternating copolymers can be obtained when MMA and St mixture is irradiated at low temperature? Or what are the conditions for aggregation (or association) of the monomers?

Chapiro

We have not observed any anomaly in copolymer composition when irradiating styrene and MMA at low temperature. It is at present too early to give general rules for the formation of aggregates which lead to controlled chain propagation. Most mixtures of polar molecules will form aggregates at certain temperatures. As stated in the paper, the fact that mixture of crystallisable compounds form glasses at low temperature is a strong indication for the existence of plurimolecular aggregates, but these may or may not lead to controlled chain propagation.

ESR Studies of Radiation- and Photo-Induced Ionic Polymerizations

By H. Yoshida, M. Irie and K. Hayashi

Faculty of Engineering, Hokkaido University, Sapporo, Japan

Introduction

Propagating chain ends in ionic polymerization carry a carbonium ion, oxonium ion, carbanion and so forth, which are generally undetectable by electron spin resonance (ESR). However, ESR still can give useful information about the initiation processes of the ionic polymerization, especially when it is used in combination with conventional chemical analysis of product polymer and some other spectroscopic methods. It is because radical ions are sometimes expected to be precursors of the propagating ions. The present authors studied ESR spectra of ion radicals formed from styrene (Iwamoto et al., 1969; Tsuji et al., 1969), α-methylstyrene (Yoshida et al., 1969), vinylethers (Irie et al., 1969) and nitroethylene (Tsuji et al., 1967) in gamma-irradiated organic glassy matrices containing a small amount of these monomers. These observations gave an insight into the radiation-induced ionic polymerizations (Yoshida & Hayashi, 1969).

In this paper, the attention will be focussed on styrene and α-methylstyrene which are known to be polymerizable in any one of the cationic, anionic and (for styrene) radical mechanisms. The radiation-induced polymerization of styrene was studied extensively by our group and it was concluded, from polymerization kinetics, the effect of additives and copolymerization studies, that the polymerization proceeds in the cationic mechanism when styrene is rigorously dried and that the propagating chain ends are free from counter anions (Ueno, Hayashi & Okamura, 1965; Ueno, Hayashi & Okamura, 1966; Ueno et al., 1967). Independently, Potter et al. reached the same conclusion (Potter, Retton & Metz, 1966). The radiation-induced polymerization of α-methylstyrene was studied by Hubmann et al. (Hubmann, Taylor & Williams, 1966) who found that the polymerization proceeded in the cationic propagation. On the contrary, Katayama et al. reported the radiation-induced anionic polymerization of α-methylstyrene and they suggested that the anionic entities observed by the pulse radiolysis technique were involved in the initiation process of the polymerization (Katayama et al., 1965; Katayama, 1965).

Based on the concept of the radical cation as an initiating entity of the cationic polymerization, the study was extended to the photo-induced polymerization of α-methylstyrene in the presence of an electron acceptor (Irie, Tomimoto & Hayashi, 1970). α-Methylstyrene which is a weak electron donor was chosen because of the absence of its radical polymerization. It was expected that the electron donor-acceptor (EDA) complex may dissociate into the acceptor anion and the α-methylstyrene radical cation in polar solvent under photo-irradiation and that the cationic polymerization may be initiated by the radical cation. Eventually, this was confirmed by the effect of additives and the reactivity ratios in copolymerization with styrene (Irie, Tomimoto & Hayashi, 1970). The photo-dissociation of the EDA complex was studied, in connection with the initiation process of the polymerization, by ESR as well as optical absorption and emission spectroscopy and electric conductivity measurements (Irie, Tomimoto & Hayashi, 1972).

It will be surveyed below how the ESR technique contributed to and what are the limitations of this technique in the studies of radiation- and photo-induced cationic polymerizations of styrene and α-methylstyrene.

Experimental

All chemicals were purified and dried very carefully as described before (Iwamoto et al., 1969; Yoshida et al., 1969; Ueno, Hayashi & Okamura, 1965; Irie, Tomimoto & Hayashi, 1970). Styrene and α-methylstyrene were washed with aqueous solution of sodium hydroxide, dried with calcium hydride and then distilled in a vacuum system onto barium oxide baked beforehand. Sample solutions were sealed in quartz ESR tubes under a high vacuum, except otherwise stated. Gamma-irradiation was carried out with a ^{60}Co source at 77 K in dark. For photo-irradiation, the light from a 500 W Xe lamp (Ushio, UXL-500) was used through a cut-off filter.

ESR measurements were carried out with a conventional X-band spectrometer with a 100 kHz magnetic field modulation. All gamma-irradiated samples were measured at 77 K in the dark. In the study of the photo-induced polymerization, the samples were kept in the resonant cavity and measured during and after the photo-irradiation through a light port of the cavity. Temperature of the samples was usually somewhat lowered to increase the spectral intensity. To study short-lived free radicals during photo-irradiation at room temperature, the sample solution in the ESR flat cell was exchanged by flowing slowly after purging with helium gas to avoid the depletion of the concentration of solutes and an intense light source (Philips, SP-500) was used. Details are described elsewhere (Yoshida & Warashina, 1971).

Results and Discussion

Radiation-Induced Polymerization

2-Methyltetrahydrofuran glass irradiated by gamma-rays at 77 K is known to give an ESR signal composed of a sharp single line spectrum due to trapped electron and a seven-line spectrum due to free radicals formed from glass matrix molecules. In the presence of a small amount (about 1 mol-%) of styrene, the trapped electron spectrum is replaced with a new single line spectrum with a total spread of 45 G. This is attributed to the styrene radical anion formed by the electron capture reaction of styrene.

When *n*-butylchloride glass is irradiated with gamma-rays at 77 K, a six-line spectrum is observed which is attributed to *n*-butyl radical. The primarily formed electron reacts readily with a butylchloride molecule and therefore is not trapped in this glass. In the presence of styrene as solute, a broad single line spectrum with a total spread of 55 G is additionally observed. It is attributed to the styrene radical cation formed by a positive hole transfer from the glass matrix to a solute.

Although the hyperfine structure of the ESR spectra due to added styrene is not resolved because of the rigid trapping in the glasses, the observed total spreads are well accounted by a calculation of spin densities and excess charge densities with the simple LCAO-MO and the McConnell's relation modified by Colpa & Bolton (Colpa & Bolton, 1963). The estimated sum of all isotropic hyperfine constants are 37.5 and 48.8 G for the radical anion and cation, respectively, which agrees with the observed spread if one assumes a width, 7G, of each hyperfine line caused by the dipolar term (Iwamoto et al., 1969).

Similarly, the radical anion and cation of α-methylstyrene were found to be formed in the gamma-irradiated organic glassy matrices. Fig. 1 *a* shows an ESR signal observed from neat 2-methyltetrahydrofuran after the irradiation. The central sharp spectrum due to the trapped electron is replaced by a broad single line spectrum due to the radical anion in the presence of α-methylstyrene, as shown in Fig. 1 *b*. Its spectral shape is obtained by subtracting the seven-line spectrum due to the free radical formed from 2-methyltetrahydrofuran molecule from the signal of Fig. 1 *b*, which has the width (peak-to-peak) of 14 G, as shown in Fig. 1 *c* (solid line). The observed shape is in agreement with that expected theoretically for the α-methylstyrene radical anion (dotted line). On the other hand, Fig. 2 *a* shows the six-line spectrum due to *n*-butyl radical formed in the neat *n*-butylchloride glass. In the presence of α-methylstyrene, a broad single line spectrum due to the radical cation superposes on the six-line spectrum as shown in Fig. 2 *b*. The observed spectral shape of the radical cation with the width of 35 G agrees well with that theoretically expected, as shown in Fig. 2 *c* (Yoshida et al., 1969).

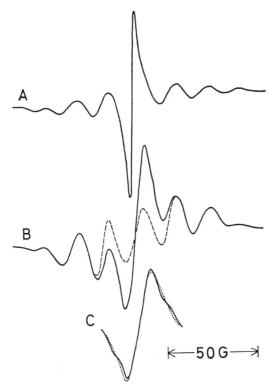

A

B

C

$\longleftarrow$50 G$\longrightarrow$

Fig. 1. ESR spectra of 2-methyltetrahydrofuran glasses irradiated by γ rays to the dose of 3.5×10^{18} eV/g at 77 K. *A*, Neat glass; *B*, glass containing 5.31 mol-% α-methylstyrene before (solid line) and after (dashed line) photobleaching; *C*, spectrum due to added α-methylstyrene, observed shape (solid line) and expected shape simulated from the MO calculation (dotted line).

These results indicate that both styrene and α-methylstyrene have the nature to form either of radical anion and radical cation. Furthermore, the observation that the yield of the radical ions reaches to a plateau value even at the low solute concentration of 0.1–1 mol-% indicates a high efficiency of the formation reaction of the radical ions. A small concentration of the monomers captures all electrons or positive holes formed in the glasses by gamma-rays. In the plateau region, the yield of the radical anions in 2-methyltetrahydrofuran and that of the radical cations in *n*-butylchloride are about 2 and 1, respectively.

Attempts were made to obtain, by the pulse radiolysis, a direct evidence of cationic entities responsible for the radiation-induced cationic polymerization of styrene (Metz, Potter & Thomas, 1967) and α-methylstyrene, but failed. It was only reported by Shida & Hamill that the 2-methyltetrahydrofuran and butylchloride glasses containing styrene showed optical absorption bands at 410 and 350 nm, respectively, attributable to the styrene radical anion and cation, after the gamma-irradiation at 77 K. The present ESR results confirm this assignment of the absorption bands.

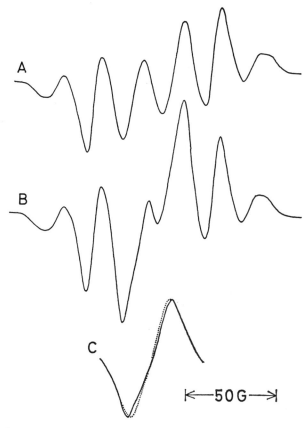

Fig. 2. ESR spectrum of *n*-butyl chloride glasses irradiated by γ rays to the dose of 3.5×10^{18} eV/g at 77 K. *A*, Neat glass; *B*, glass containing 0.69 mol-% α-methylstyrene; *C*, spectrum due to added α-methylstyrene, observed shape (solid line) and expected shape simulated from the MO calculation (dotted line).

Polystyrene was obtained only from *n*-butylchloride glass but not from 2-methyltetrahydrofuran glass. This shows the correlation between the radical cation and the cationic polymerization. It might be essential to study the reaction of the radical cation. However, we could not elucidate it by ESR, because the change of the radical cation spectrum was not clear because of the lack of hyperfine structure and the overlapping of the *n*-butyl radical spectrum. On the other hand, the radical anion of styrene (Tsuji et al., 1969) and α-methylstyrene (Lin, Tsuji & Williams, 1968) was found to be readily protonated in organic glasses. This gives one of the possible reasons why the anionic polymerization was not important in the radiation-induced polymerization of these monomers.

Based on the above results, the pulse radiolysis study of styrene in 2-methyltetrahydrofuran and *n*-butylchloride was made at room temperature. Immediately after the electron pulses, the absorption bands due to the radical anion and cation were observed at 390 and 350 nm. The latter disappeared by a second order decay at the early period when the concentration of the radical cation and anionic entities is high enough. The decay deviated from the second order curve, as the concentration decreased. This can be reasonably attributed to that the first order decay due to the reaction between the radical cation and styrene monomer becomes more and more pronounced. The rate constant was estimated to be 2×10^6 M^{-1} s^{-1} for this reaction (Yoshida, Noda & Irie, 1971).

After all, these monomer molecules efficiently capture electron and positive hole primarily formed by gamma-rays, so that their radical anion and cation are formed as secondary products. The former is readily protonated, while the latter adds to a monomer molecule and forms a carbonium ion responsible for the cationic propagation.

Photo-Induced Polymerization

Recently, Irie et al. studied the photo-induced polymerization of α-methyl-styrene in polar solvent such as 1,2-dichloroethane in the presence of tetra-cyanobenzene and proved that it proceeds in cationic propagation (Irie, To-mimoto & Hayashi, 1970). Tetracyanobenzene is a typical electron acceptor, while α-methylstyrene is thought to be a weak electron donor. As a matter of fact, the formation of the EDA complex between them was evidenced by observing a charge transfer band with a maximum at 363 nm as well as an acceptor band with a maximum at 316 nm (Irie, Tomimoto & Hayashi, 1972).

The ESR spectrum observed from a viscous solution of tetracyanobenzene and α-methylstyrene in amyl alcohol at 223 K during photo-irradiation is composed of equally spaced nine hyperfine lines, which is attributed to four equivalent nitrogen nuclei in tetracyanobenzene anion. When the spectrum is recorded at room temperature from *n*-butanol solution flowing through the ESR flat cell, the width of each line becomes much narrower, to less than 0.1 G, which enables us to determine the isotropic hyperfine constant of 1.04 G, as shown in Fig. 3. The steady state concentration and the lifetime observed at room temperature were about 10^{-7} M and less than 0.1s.

These results indicate that the EDA complex dissociates from its excited state to ions, tetracyanobenzene anion and necessarily the α-methylstyrene radical cation. The latter was not observed by ESR, though its formation was evidenced by photo-irradiating *n*-amyl alcohol glass containing both tetra-cyanobenzene and α-methylstyrene and observing optical absorption bands due to the radical anion at 462 nm and dimer radical cation around 480 nm, which resulted from the α-methylstyrene radical cation, simultaneously. The reason why the ESR spectrum of the α-methylstyrene radical cation was un-observable is still unknown.

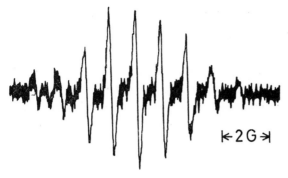

Fig. 3. ESR spectrum observed from *n*-butanol solution containing 1.5 M of α-methylstyrene and 6×10^{-4} M of tetracyanobenzene during photoirradiation at 298 K.

The ESR spectrum of the radical anion of tetracyanobenzene is observed with the light of wavelength longer than 350 nm which covers only the charge transfer band. The intensity of the spectrum increases for the light of $\lambda > 290$ nm covering both the acceptor band and the charge transfer band. The transient rise of the concentration of radical anion on turning on the light follows an S-shaped curve, as shown in Fig. 4, suggesting a long life of the precursor of the radical anion. These results imply that the EDA complex in the singlet-excited state is formed either by the photoexcitation of the complex in the ground state or by the photoexcitation of the acceptor followed by a heteroexcimer formation and that the excited EDA complex dissociates into ions through a triplet excited state.

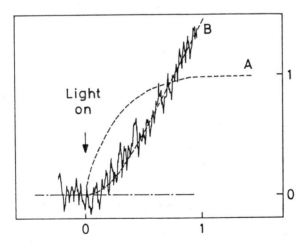

Fig. 4. *Abscissa:* time (s); *ordinate:* (*left*) ESR signal intensity; (*right*) conc. triplet (arb).
Transient rise of the intensity of ESR spectrum during photoirradiation of *n*-amyl alcohol solution containing 7.6×10^{-1} M of α-methylstyrene and 1×10^{-2} M of tetracyanobenzene at 123 K when the light is turned on. Curves A and B show, for reference, the accumulation of triplet excited EDA complex and that of the tetracyanobenzene anion estimated from the life-time of the triplet excited EDA complex determined by phosphorescence measurements.

By means of the fluorescence and phosphorescence spectroscopy, the flash photolysis and the flash electric conductivity measurements, the results obtained were consistent with the above mechanism of the formation of the tetracyanobenzene anion and the α-methylstyrene radical cation. However, it should be noted that a triplet quencher, O_2, did not eliminate the ESR spectrum of the radical anion completely. This suggests a possibility that the singlet-excited EDA complex also dissociates directly into ions (Irie, Tomimoto & Hayashi, 1972).

The photo-induced cationic polymerization was also found in pyromellitic dianhydride-α-methylstyrene system. The ESR spectrum with three hyperfine lines due to the radical anion of pyromellitic dianhydride was observed from the same system during the photo-irradiation. Thus, the ESR studies strongly suggest that the α-methylstyrene radical cation is the primary entity in the photo-induced cationic polymerization in the presence of an electron acceptor.

References

Colpa, J. P. & Bolton, J. R., Mol. Phys., *6*, 273 (1963).

Hubmann, E., Taylor, R. B. & Williams, Ff., Trans. Faraday Soc., *62*, 88 (1966).

Irie, M., Hayashi, K., Okamura, S. & Yoshida, H., Int. J. Rad. Phys. Chem., *1*, 297 (1969).

Irie, M., Tomimoto, S. & Hayashi, K., J. Polymer Sci., *8*, 585 (1970).

Irie, M., Tomimoto, S. & Hayashi, K., J. Phys. Chem., *76*, 1419 (1972).

Iwamoto, T., Hayashi, K., Okamura, S., Hayashi, Ka. & Yoshida, H., Int. J. Rad. Phys. Chem., *1*, 1 (1969).

Katayama, M., Hatada, M., Hirota, K., Yamazaki, H. & Ozawa, Y., Bull. Chem. Soc. Japan, *38*, 851 (1965).

Katayama, M., Bull. Chem. Soc. Japan, *38*, 2208 (1965).

Lin, J., Tsuji, K. & Williams, Ff., Trans. Faraday Soc., *64*, 2896 (1968).

Metz, D. J., Potter, R. C. & Thomas, J. K., J. Polymer Sci., A-1, *5*, 877 (1967).

Potter, R. C., Retton, R. H. & Metz, D. J., J. Polymer Sci., A-1, *5*, 2295 (1966).

Tsuji, K., Iwamoto, T., Yoshida, H., Hayashi, K. & Okamura, S., Memoirs Fac. Engineering, Kyoto Univ., *31*, 268 (1969).

Tsuji, K., Yamaoka, H., Hayashi, K., Kamiyama, H. & Yoshida, H., J. Polymer Sci., *B4*, 629 (1967).

Ueno, K., Hayashi, K. & Okamura, S., J. Polymer Sci., *B3*, 363 (1965).

Ueno, K., Hayashi, K. & Okamura, S., Polymer, *7*, 431 (1966).

Ueno, K., Williams, Ff., Hayashi, K. & Okamura, S., Trans. Faraday Soc., *63*, 1478 (1967).

Yoshida, H., Hashimoto, S., Iwamoto, T. & Okamura, S., Bull. Chem. Res. Kyoto Univ., *47*, 1 (1969).

Yoshida, H. & Hayashi, K., Adv. Polymer Sci., *6*, 401 (1969).

Yoshida, H. & Warashina, T., Bull. Chem. Soc. Japan, *44*, 2950 (1971).

Yoshida, H., Noda, M. & Irie, M., Polymer J., *2*, 359 (1971).

Discussion

Williams

From your pulse radiolysis experiments on the decay of the styrene radical cation, are you able to estimate an upper limit for the rate constant of the bimolecular reaction between the radical cation and the monomer?

Yoshida

We obtained the value of 2×10^6 M^{-1} s^{-1} for the bimolecular rate constant as mentioned in the text, which gives only an estimate of upper limit. The extinction coefficient of the styrene radical cation (1.6×10^3 M^{-1} cm^{-1} at 333 nm) used seems to be too small. If one takes a more probable value of 10^4 M^{-1} cm^{-1}, the estimate of the rate constant becomes as small as 10^5 M^{-1} s^{-1}.

CROSSLINKING REACTIONS

Free Radicals Observed during the Cure of Unsaturated Polyester Resins

By Pekka Lehmus[1]

Institute of Polymer Technology, Royal Institute of Technology, Stockholm, Sweden

The curing process of unsaturated polyester resins can be described as a free radical copolymerization reaction between polyester fumarate units and styrene monomer. Due to the occurrence of several double bonds per polyester molecule, the polymerization proceeds with extensive crosslinking and results in three-dimensional network. This structure determines the characteristics of the curing process. The gelation occurs very early during the cure causing suppression of termination reactions, which results in great acceleration of curing reactions and temperature increase. This is useful in practice, because the cure is often accomplished without external heating using redox-type initiation. As a rule, such a cure is not complete, as can be directly observed by IR analysis of residual unsaturation (Imai, 1967).

It is known, however, that curing reactions slowly continue even at room temperature, as can be confirmed by the IR method or through the change of the mechanical properties of the resin (Learmont, Tomilson & Czerski, 1968). One interesting question concerning late curing reactions is the importance of the free radicals observed by means of ESR at this stage of the cure in polyester resins (Demmler & Schlag, 1967). The simplest assumption would be that the radicals observed are propagating radicals of styrene and fumarate type, and the ESR spectrum would give direct indication of the rate of the curing reactions. Other possibilities are that the observed radicals are occluded radicals which have no possibility to take part in further polymerization reactions or secondary radicals caused by various chain transfer reactions of the primary propagating radicals.

Experimental

Polyester resins were kindly provided by SOAB and were of fumarate-phthalate-propylene glycol type with varying proportions of dibasic acids. No difference in ESR spectra was observed, as in some experiments resins prepared

---

[1] Present address: Neste Oy, Tutkimuskeskus, Kulloo, Finland.

in the laboratory from carefully purified chemicals were used. Initiators and accelerators, cyclohexanon peroxide, benzoyl peroxide, cobalt naphthenate and dimethyl aniline, were of commercial quality and were used without further purification.

Samples for ESR were prepared in 3 mm inner diameter quartz tubes, which were rapidly evacuated after mixing the components. ESR spectra were taken with JEOL JES-3B spectrometer equipped with variable temperature apparatus.

Results and Discussion

A typical ESR spectrum for a polyester resin after redox-initiated room temperature cure is shown in Fig. 1*a*. The asymmetric shape of the spectrum is hardly due to any anisotropy effects but is caused by the overlapping of two spectra with slightly different *g* values, as is roughly shown in Fig. 1*b*. Heating of the sample at 80°C or above causes a notable increase in the intensity of the spectrum while simultaneously the original hyperfine structure gradually disappears (Fig. 2). If the heating is continued, the spectrum slowly disappears. It has been shown (Demmler & Schlag, 1967) that the increase of the total intensity depends on the amount of peroxide initiator added to the resin and is ultimately due to the further reaction of the residual peroxide still present in the resin after the cure. The presence of the peroxide after the cure has been confirmed in an earlier study (Alt, 1962).

The broad three-line spectrum observed immediately after the room temperature cure is similar to the spectrum observed in irradiated polystyrene and in polystyrene mechanically degraded in vacuo (Bresler et al., 1959). In these cases the spectrum is supposedly caused by tertiary styryl radicals, $-CH_2-\dot{C}(Ph)-CH_2-$ (Tino, Capla & Szöcs, 1970). In polyester resin system, these would be typical chain transfer radicals which eventually could cause branching of polystyrene chains. To observe the propagating type radicals independently, trials were made with the rapid-flow system developed for this purpose (Dixon & Norman, 1963). $TiCl_3/H_2O_2$ initiation system was used in acidified methanol solution. ESR spectra recorded for maleate and fumarate esters and for low molecular weight polyester are shown in Fig. 3. Furthermore, a trial to record a corresponding spectrum for styrene is represented in Fig. 4. Coupling constants for styrene monomer radicals were estimated in analogy with the published coupling constants for benzyl radical (Carrington & Smith, 1965). To observe the solid-state appearance of the spectra, some experiments were carried out with model substances in frozen state at liquid nitrogen temperature. For diethyl fumarate, a broad two-line spectrum was obtained, Fig. 5, while the spectrum for styrene was a very broad one-line spectrum, consisting probably of several overlapping spectra. Through comparison of

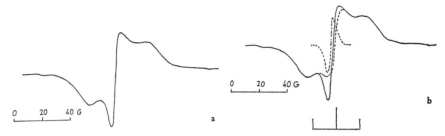

Fig. 1 *a*, Typical spectrum for polyester resin after room temperature cure (no heat treatment) Polyester Soredur H 65 with 2 % Cyclonox B-60 peroxid and 0.012 % cobalt naphthenate. *b*, Same spectrum divided in two components, ----, $g = 2.004$; ——, $g = 2.002$.

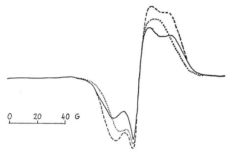

Fig. 2. ESR spectra during heat treatment at 80°C. Same polyester resin as above, ——, 20 min; ---, 60 min; ......, 180 min.

the recorded spectra it is possible to propose that polyester resin, initially after the room temperature cure, contains free radicals of the fumarate propagating and styryl tertiary radical types.

It is possible to cite some further observations, which seem to strengthen this proposition. Considerable branching was observed in styrene-fumarate co-

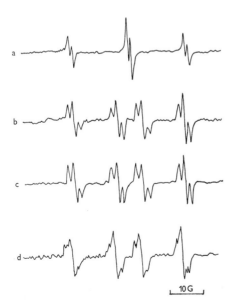

Fig. 3. ESR spectra recorded with rapid flow system. $TiCl_3/H_2O_2$ initiation system in acidified methanol. *a*, Initiation system only; *b*, diethyl maleate; *c*, diethyl fumarate; *d*, poly(ethylene glycol) fumarate).

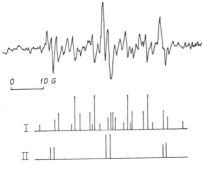

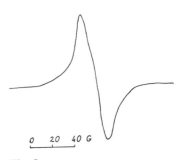

Fig. 4   Fig. 5

Fig. 4. ESR spectrum of styrene radical by using rapid flow system $TiCl_3/H_2O_2$ initiation system in acidified methanol, styrene concentration 0.5 mol l^{-1}. I, HO–CH$_2$–CH$_2$–ĊH(Ph); II, HO–ĊH$_2$.

Fig. 5. ESR spectrum of diethyl fumarate with 0.1 % benzoyl peroxide after 120 min UV irradiation at $-196°C$.

polymer chains, as thermally postcured resins were hydrolyzed (Feinauer, 1963). For such branching, tertiary styryl radicals have to occur at some stage of the curing process. Further, IR analysis of the residual unsaturation has shown that polyester double bonds are often sterically hindered and fail to react in a higher proportion than styrene double bonds (Demmler & Ropte, 1968). Finally, it must be mentioned that recently published results (Demmler & Schlag, 1971) seem to confirm the character of the observed free radical concentration as a dynamic steady-state concentration, where new free radicals are continuously generated.

Therefore, it seems reasonable to suppose that the observed free radicals play a significant role in the late stages of the curing reactions. It is, however, on the basis of the present knowledge not possible to analyze the role of the radical (or radicals) formed during the heat treatment, giving the stable one-line spectrum.

This work has been financed by grants from The Swedish Board for Technical Development and through a scholarship from Neste Foundation, Finland.

References

Alt, B., Kunststoffe, *52*, 133 (1962).
Bresler, S. E., Jurkov, S. N., Kazbekov, E. N., Saminsky, E. M. & Tomashevsky, E. E., J. Techn. Fiz. (USSR), *29*, 359 (1959).
Carrington, A. & Smith, I. C. P., Mol. Phys., *9*, 137 (1965).
Demmler, K. & Schlag, J., Kunststoffe, *57*, 566 (1967).
Demmler, K. & Ropte, E., Kunststoffe, *58*, 925 (1968).

Demmler, K. & Schlag, J., Farbe Lack, *77*, 224 (1971).
Dixon, W. T. & Norman, R. O. C., J. Chem. Soc., *1963*, 3119.
Feinauer, R., Dissertation, TH Stuttgart, 1963, 64, 94.
Imai, T., J. Appl. Pol. Sci., *11*, 1055 (1967).
Learmonth, G. S., Tomilson, F. M. & Czerski, J., J. Appl. Pol. Sci., *12*, 403 (1968).
Tino, J., Capla, M. & Szöcs, F., Eur. Polym. J., *6*, 397 (1970).

Enhanced Crosslinking with Acrylics

By A. Charlesby and D. Campbell

Royal Military College of Science, Shrivenham, Swindon, Wilts, UK

Exposure of polymers to high energy irradiation is known to cause crosslinking or degradation (main chain scission) according to the type of polymer. Several mechanisms have been suggested involving either radicals or ions. Present evidence favours the radical hypothesis, such evidence arising from protection by radical scavengers, ESR studies of radicals, temperature dependence etc. In particular, the formation of a crosslink requires the presence of two neighbouring radicals, one on each polymer chain.

The relationship between crosslinked density and radiation dose has been studied by various methods including solubility and elastic modulus measurements and the density of crosslinks appears to correlate with the radical concentration. However, this concentration and the doses usually employed would rarely allow two radicals to be produced independently by radiation events yet in close proximity. It is therefore necessary to postulate that in solid polymers radicals can move and decay slowly. Such mobility can be achieved by a series of hydrogen abstraction and addition reactions.

Assuming a random distribution of crosslinks, the doses required for the formation of a crosslink network can be readily evaluated. The minimum dose r_{gel} is related to the weight average molecular weight M_w by the formula

$$r_{gel} = 0.96 \times 10^6 / M_w G(R\cdot)$$

where $G(R\cdot)$ represents the number of crosslinking radicals produced per 100 eV absorbed. For polyethylene doses of about 5 or 10 Mrad are usually needed to produce a satisfactory degree of crosslinking.

A different situation arises when crosslinks occur via a chain reaction, involving both polymer and monomer. An example is the curing of unsaturated polyester/styrene mixtures. Here the initiation of a chain reaction via a single radical allows a number of crosslinks to occur in sequence. The crosslinks are therefore no longer distributed at random, furthermore, the dose requirements are considerably reduced. For this purpose one double bond per monomer unit is needed to propagate the chain and in addition one double bond per crosslink to allow the growing polymer chain to attach itself to the polyester. This extra double bond can either occur in the monomer (as in allylics) or on the polymer chain. Thus the average overall number of unsaturated groups per monomer reacted must exceed unity. Not only are the dose requirements considerably reduced, but the relationship between solubility and crosslinked density is greatly modified.

Recently a type of enhanced crosslinking has also been observed with acrylic acid (AA) and methacrylic acid (MAA) irradiated in the presence of polymer molecules (Charlesby & Fydelor, 1972). Very small doses, considerably less than 1 Mrad, are needed to achieve crosslinked networks in spite of the fact that these polymers may be fully saturated. This is a strong indication of a chain reaction proceeding via the AA or MAA in spite of the fact that these monomers only contain one double bond, and few, if any, are available on the polymer. Although the quantitative results are not very consistent, this may be explained in terms of diffusion control and other conditions prevailing during the radiation exposure. The results are nevertheless very striking.

Only certain polymers show this enhanced crosslinking e.g. polyethylene and polypropylene. The latter becomes crosslinked with AA under radiation although in the absence of acrylic acid and with higher doses it suffers simultaneous crosslinking and degradation. Irradiation of a copolymer poly-ethylene-acrylic acid does not undergo enhanced crosslinking; the acrylic acid must be present in the form of monomer.

Several attempts have been made to determine the mechanism of this reaction, e.g. by the use of suitable scavengers. Electron spin resonance studies may be expected to provide further information on the intermediates and help to elucidate the nature of the reaction. This paper describes some of this work which has provided some unexpected results even if it has not yet led to very definite conclusions.

Irradiation and Measurement at Room Temperature

The ESR spectrum of low density polyethylene, gamma irradiated at room temperature in a solution of acrylic acid in benzene or in alkane, shows the radical chain ends of polyacrylic acid, similar to those obtained from the irradiated solution alone. In addition there is a small contribution from the alkyl and allyl radicals obtained from the irradiated polyethylene alone. The radicals formed on the polyacrylic acid chains are very stable, most being retained after a month at room temperature, and this must be ascribed to radicals trapped within the polymer matrix. Thus when the acrylic acid plus polyethylene is irradiated in ethanol, no evidence for such polyacrylic acid radicals is obtained, since ethanol dissolves polyacrylic acid, and allows these radicals to react with each other. The short lifetime of these radicals also greatly re-

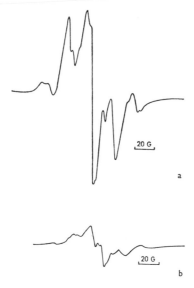

20 G

a

20 G

b

Fig. 1. ESR spectrum from (*a*) acrylic acid; (*b*) a 10 % solution of acrylic acid in benzene. γ irradiated and observed at 77 K.

duces grafting and crosslinking in polyethylene in the presence of acrylic acid in ethanol solution.

Similar results are obtained with polyethylene irradiated at room temperature in the presence of methacrylic acid solutions.

These results indicate that there are no surprising features in these room temperature ESR measurements, able to account for the enhanced crosslinking. It was therefore decided to extend these experiments to low temperatures, both for irradiation and measurement, to detect any unusual precursors for the unexpected reaction.

Low Temperature Irradiation and Measurement

The ESR spectrum of low density polyethylene samples irradiated and measured at 77 K, shows, in addition to the usual alkyl radicals, some contribution from trapped electrons and possibly positive ions. When low polyethylene is irradiated in a frozen solution of AA (10 % in benzene) there appears in addition a triplet ($a_H = 14$ G) with substructure, and a narrow central singlet. These may be bleached by infrared light or by thermal annealing. The singlet is best ascribed to trapped electrons, and the triplet to a cationic species.

The singlet and triplet do not appear from irradiation of the monomer alone, of the monomer in solution, or of the polyethylene and solvent. On the other hand it does appear with polyethylene plus pure AA. Again it does not appear if a block or grafted copolymer of ethylene and AA is irradiated at this temperature, unless AA monomer is simultaneously present.

Irradiation at 77 K of polyethylene with a corresponding solution of MAA in benzene gives a similar triplet spectrum—though not if the monomer solu-

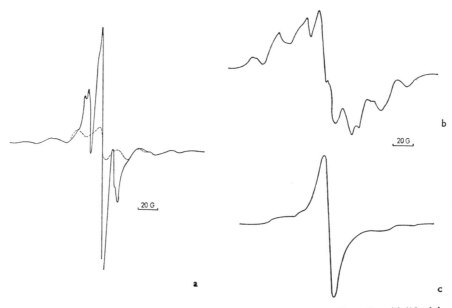

Fig. 2. ESR spectrum of (*a*) low density polyethylene in a solution of acrylic acid (10 %) in benzene, irradiated and observed at 77 K. (Note effect of exposure to light -----). This spectrum is quite different from that of irradiated acrylic acid (fig. 1); (*b*) low density polyethylene (LDPE); (*c*) an irradiated copolymer of ethylene and acrylic acid (with no monomer present).

tion is irradiated in the absence of polyethylene. Irradiated MAA solutions in the absence of polymer of course give a very different spectrum to that from AA solutions, showing the important role played by the polymer. This is further shown by the fact that AA solutions irradiated at 77 K with deuterated polyethylene shows only a small triplet part of the spectrum.

From these results it may be inferred that the triplet plus singlet part of the spectrum is due to some interaction of monomer (AA or MMA) with polymer, and even that the species responsible for the triplet is more closely associated with base polymer than with the monomer. However the acrylic or methacrylic acid must be present within the polymer as monomer copolymerised during irradiation; polyacrylic acid copolymerized with polyethylene does not produce the effect, nor does it show enhanced crosslinking.

Other Base Polymers

When the low density polyethylene is replaced by high density polyethylene, polypropylene, poly(ethylene-acrylic acid) or poly(3-methyl pentene), and irradiated in the presence of a solution of acrylic acid monomer, a crosslinked network is formed at very low doses. When the irradiation is carried out at 77 K, the triplet part of the spectrum is clearly observed. Conversely poly-

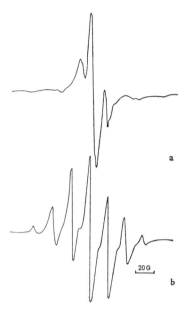

a

b

20 G

Fig. 3. (*a*) ESR spectrum of LDPE irradiated at 77 K in presence of a 10 % solution of methacrylic acid in benzene. The spectrum is similar to that with acrylic acid solution (Fig. 2) and quite different from that of LDPE (Fig. 2) or (*b*) the monomer itself.

ethylene glycol and polymethyl methacrylate do not exhibit enhanced cross-linking when irradiated in the presence of AA, nor do they show the triplet spectrum at 77 K. Polyethylene terephthalate, polytetrafluorethylene and polyvinyl chloride plus AA, irradiated at 77 K, do not show the triplet spectrum, for which it appears that a hydrogen-containing polymer is a necessary though not sufficient condition (cf. deuterated polyethylene).

Nylon 6 and Nylon 11 do give a triplet spectrum with AA, but this is not similar to that seen with polyethylene + AA. In any case nylon does not show enhanced crosslinking.

Other Vinyl Monomers

Of the various vinyl monomers examined, only AA and MAA show the ability to produce enhanced crosslinking in polymers, notably polyethylene and polypropylene. It is therefore important to compare their ESR spectra with those of other vinyl monomers.

Ethyl acrylate, *n*-butyl acrylate and methyl methacrylate show triplets on irradiation at 77 K, but only in the presence of the polymer. The coupling constants are somewhat lower than for AA and MAA and these monomers do not promote crosslinking. Styrene, acrylonitrile and allyl alcohol irradiated in the presence of polyethylene give a broad singlet with no sign of the triplet. Vinyl acetic acid, allyl acetate and vinyl acetate treated in the same way show intermediate spectra, often with a quartet.

Table 1. *ESR Spectrum of AA and related Monomers*

Low density polyethylene + monomer (10 % in benzene) irradiated at 77 K

		ESR spectrum includes
CH$_2$ = CH | COOH Acrylic acid	CH$_2$ = C(CH$_3$) | COOH Methacrylic acid	Triplet
CH$_3$—CH = CH | COOH Crotonic acid	CH$_3$—CH = C(CH$_3$) | COOH Tiglic acid	Quintet
(CH$_3$)$_2$—C = CH | COOH 3,3-Dimethacrylic acid		Septet

Monomers Related to Acrylic Acid

Taking acrylic acid as standard, we have examined related monomers, irradiated at 77 K in 10 % solutions in benzene in the presence of low density polyethylene. The ESR results can be summarised in Table 1. In the absence of the polymer very different spectra are obtained. This would indicate that the spectra arise from an unpaired electron interacting with groups attached to the terminal C atom (Campbell & Charlesby, 1971). Substitution in position 2 has little effect. These observations appear to favour the monomer as the origin of the triplet, a result in apparent contradiction with that derived from the examination of the various polymers discussed above, e.g. the effect of deuteration in polyethylene. One possibility is a resonance.

CH$_2$=CH $\rightleftarrows$ ĊH$_2$CH $\rightleftarrows$ CH$_2$=CH
$\quad$ \| $\qquad\qquad$ \|\| $\qquad\qquad\quad$ \|
$\quad$ C· $\qquad\qquad$ C $\qquad\qquad\quad$ C
$\quad$ / \ $\qquad\quad$ / \ $\qquad\quad$ / \\
O $\quad$ O$^+$ $\quad$ O $\quad$ O$^-$ $\quad$ $^+$O $\quad$ O
H $\qquad\quad$ H $\qquad\qquad$ H

Crystalline and Amorphous Low Molecular Polymers

It has already been reported that linear paraffins can be incorporated in a growing polyacrylic chain, to an extent far greater than can be expected from the number of radicals formed directly by radiation on the paraffin. This is the analogue of enhanced crosslinking of polyethylene; the solubility of the system is merely due to the short length of the paraffin, so that it is unlikely for a single molecule to be attached to more than one poly acrylic chain. One may therefore enquire whether such mixtures, when irradiated at 77 K, show a triplet structure in their ESR spectra.

10 % solutions of AA in *n*-hexane or in *n*-hexadecane irradiated at 77 K give ESR spectra typical of the irradiated paraffin. This could be explained as due to separation of the components on freezing. Higher molecular weight paraffins (dotriacontane and hexatriacontane) irradiated in the presence of a 10 % AA solution in benzene do show an ESR spectrum with a triplet component. Squalane, a branched paraffin which forms a glass on cooling, does give the typical triplet when irradiated at 77 K in the presence of this AA solution; however 3-methylpentane, which also forms a glass gives no evidence of the triplet under similar irradiation conditions with AA solution.

Discussion

Many measurements of enhanced crosslinking have been made, using solubility to determine the crosslink density. Results are not always as consistent as might be desired, but do emphasize that a chain reaction of high efficiency is involved. Two requirements appear essential; the presence of hydrogen atoms in the polymer; the simultaneous presence of AA or MAA monomer during irradiation with their ability to polymerise and to contribute COOH groups. (Polyacrylic acid does not meet the polymerisable requirement.)

Enhanced crosslinking can be expected in systems where each bridge is associated with more than enough double bonds to form a linear chain. Examples are allylic monomers (with two double bonds per monomer) and unsaturated polyester-styrene mixtures. AA and MAA meet this requirement if we allow the COOH groups to participate in the reaction, forming additional links. This is best achieved by some form of radical-ion reaction, induced by radiation, and several tentative explanations have been advanced. These will be discussed elsewhere. Here it seems more appropriate to consider whether electron spin resonance can provide evidence for the nature of the radical-ion. The results summarised here favour association with the triplet spectrum, but the evidence is not conclusive. Certain discrepancies remain to be resolved; it may be that the radical-ion is only one of the necessary conditions satisfied in AA and MAA, but that other necessary conditions are not met in certain polymers. Although the location of the radical-ion appears to be on the monomer, the evidence indicates that it is also associated with the base polymer whose presence is essential. The nature of this association offers an intriguing challenge to electron spin resonance research. A further feature is the ability to bleach the species involved with light or heat, and this offers interesting possibilities of following the reaction by the effect of this bleaching on both enhanced crosslinking and the ESR spectrum, as well as on the optical absorption and possibly even on the electrical properties of the irradiated polymer-monomer mix.

References

Campbell, D. & Charlesby, A., Chem. Zvesti, *26*, 253 (1972). (Proc. Int. Conf. on Chemical Transformations of Polymers, Bratislava, June (1971)).
Campbell, D. & Charlesby, A., Eur. Polymer J. In press.
Charlesby, A. & Fydelor, P. J., Int. J. Radiat. Phys. Chem., *4*, 107 (1972).

Discussion

Ivin

Can you rule out the possibility of simultaneous or consecutive generation of branch points on the same hexadecane molecule, followed by termination by mutual combination of two growing acrylic acid chains?

Have you tried treating the hexadecane/acrylic acid copolymer chemically to discover the nature of the chemical linkages between the hexadecane chains and the acrylic acid chains?

Charlesby

There is insufficient energy available to produce one radical per hexadecane molecule, in fact considerably less than 1% can be affected directly by the radiation. We have looked at infrared, but the expected changes would be small as compared with a simple mixture of AA and C_{16} unless some bonds of an entirely new type are formed; at the most 1 or 2% of the bonds are modified.

Chapiro

Crosslinking by a chain mechanism of a saturated polymer (such as polyethylene) can only occur if one uses a difunctional monomer. Acrylic acid is a monofunctional monomer which readily polymerizes to a linear chain molecule with pendant carboxyl groups. It follows that under irradiation of polyethylene acrylic acid mixtures a structure is expected to arise in which poly-(acrylic acid) branches are grafted to the polyethylene backbone. This results in a non-crosslinked structure. Crosslinking can occur, however, in a second stage on heating this polymer whereby polyacrylic branches could be linked to each other by anhydride bridges. Another possibility would be that the acrylic acid grafted polyethylene chains are no longer soluble in the usual solvents of polyethylene (e.g. xylene) and that the apparent "gel" is actually a non-crosslinked, insoluble graft copolymer.

Charlesby

Your first point is agreed; heating does produce a change, but not very great. But one does not find the equivalent change in the copolymer. Apparently acrylic acid must be present as monomer to achieve the effect.

At the doses used (well below 1 Mrad), there are insufficient radicals produced to give one per polyethylene molecule. If we had one radical per polymer molecule there would be no problem. Straightforward crosslinking could occur with or without crosslinking. The same remark concerns insolubility of the copolymer due to lack of a suitable solvent. Most polyethylene would not be linked into the system, and would dissolve out. In fact more than half are retained in the network after prolonged extraction in boiling xylene. Polyethylene molecules within a skin of the network, but not subjected to the AA crosslinking can be dissolved out through the crosslinked skin. Diffusion is therefore not a great hindrance.

Lindberg

Does the reaction formally belong to vulcanization reactions?

Charlesby

No. In vulcanization there is a question of a two-double bond system, here is only one double bond.

Peterlin

Does the molecular weight of the product agree with the suggested model? What is the ratio of acrylic and paraffinic component of the product? What is the length of acrylic bridge between two consecutive hexadecane molecules?

What is the intrinsic viscosity of the product? For a comb type short chain branching the value is expected to be below that of the backbone.

Charlesby

For the hexadecane–acrylic acid system, the copolymer comprises about 1.1×10^3 AA monomers + 47 hexadecane i.e. over 20 AA units between successive links. The molecular weight was measured by osmometry and estimated from the radical concentration. There were no measurements of intrinsic viscosity.

Williams

Does the crosslinking mean propagation?

Charlesby

Molecules must have two reactive sites to get the net structure.

Szwarc

One double bond is enough to get the net structure, is it not?

Charlesby

No, it cannot make the structure ($\equiv$).

PHOTOCHEMICAL DEGRADATION
AND OXIDATION

Degradation of 1,4-Polydienes Induced by Ultraviolet Irradiation

By Peter Carstensen[1]

Institute of Polymer Technology, The Royal Institute of Technology, Stockholm 70, Sweden

The effects of UV light on polymers are of practical importance, e.g. for the outdoor stability of plastics, rubbers, organic coatings, and fibres. The effects have been studied extensively by means of degradation and crosslinking experiments, gas evolution, infrared and nuclear magnetic resonance spectroscopy, etc. Although there is no doubt that free radicals are formed as primary reaction products from UV irradiation, there are few reports in the literature describing studies of these free radicals, e.g. by using ESR analysis for structural studies. Most studies of free radicals in polymers have been made with ionizing irradiation as initiators, in particular γ rays from ^{60}Co cells have been used. The γ quanta have energies of approx. 1 MeV which is several hundred thousand times higher than the energies of normal chemical bonds. UV quanta from a mercury lamp have on the other hand, strong spectral lines at the wavelengths 2537, 3130, and 3655 Å which correspond to the energies 4.88, 3.95, and 3.39 eV, respectively. Common chemical bonds in polymers, e.g. C–H and C–C, have bond strengths of 4.28 and 3.44 eV, respectively. Bond energies and UV quanta are accordingly in the same energy range. It is therefore expected that UV quanta can break chemical bonds and that the breaks will occur more selectively than with ionizing radiation. For our studies of UV initiated radicals, five polydienes with similar structures have been selected: *cis*- and *trans*-1,4-polyisoprene, *cis*-1,4-polybutadiene, *cis*-1,4-polypiperylene, and polychloroprene.

$-(CHR'-CR''\!=\!CH-CH_2)_{\overline{n}}$

$R' = R'' = H$ polybutadiene

$R' = H, \quad R'' = CH_3$ polyisoprene

$R' = CH_3, R'' = H$ polypiperylene

$R' = H, \quad R'' = Cl$ polychloroprene

All the monomer units in these polymers contain a double bond. Therefore,

[1] Present address: Chemical Research Laboratory, NKT A/S, DK-2000 Copenhagen F., Denmark.

it follows that radicals formed by simple C–C or C–H bond scissions can be of allyl, alkyl or vinyl type. By considering the energy necessary for the formation of these three different kinds of radicals, it is evident that the formation of allyl radicals requires less energy than the other types (Carstensen, 1970 *b*). Consequently it is expected that mainly allyl radical structures will be formed during UV irradiation.

Materials and Methods

Two samples of *cis*-1,4-polyisoprene (Firestone Tire and Rubber Co., B. F. Goodrich Co.) with about 95 % *cis*-1,4- and 5 % 3,4-configuration and one sample of *trans*-1,4-polyisoprene (from Polymer Corp.) with 92 % *trans*-1,4- and 2 % 1,2-configuration and one sample of *cis*-1,4-polybutadiene (B.F. Goodrich Co.) containing 98 % *cis*-1,4- and 2 % 1,2-configuration were used. Furthermore a laboratory sample of *cis*-1,4-polypiperylene (supplied by Prof. G. Natta) and a commercial sample of polychloroprene (British Geon Ltd) were investigated. All the commercial samples contained stabilizers and were accordingly purified by repeated dissolution in toluene and precipitation by addition of acetone under a nitrogen atmosphere.

For ESR measurements the samples were inserted in quartz sample tubes (5 mm outer diameter) and degassed in vacuum (better than 10^{-4} mm Hg for 48 h), and then sealed. The UV irradiation was made with a medium pressure mercury lamp (Phillips, HPK 125 W) at 77 K. The lamp emits a spectrum ranging from 2400 to 6000 Å including intense spectral lines at 2537, 2653, 2967, 3025, 3130, and 3655 Å.

The ESR spectra were recorded with an X-band spectrometer (model JES-3B from Japan Electron Optics Lab. Co.) with 100 kHz and 80 Hz field modulation. To avoid power saturation effects, the ESR measurements were carried out at a microwave power less than 1 mW. All spectra were recorded at 77 K with a 2.8 G width of field modulation. A cylindrical TE_{011} cavity was used in the present study.

Free Radicals in *cis*-1,4-Polybutadiene

In previous studies with ionizing radiation described in the literature (Kuzminskii, Fedoseva & Buchachenko, 1965; Libby, Ormerod & Charlesby, 1960; Kozlov & Tarasova, 1966) only singlet-like ESR spectra were observed. Because of this lack of hyperfine structure it was not possible to relate the spectra to any specific radical structure.

On the other hand, by UV irradiation of *cis*-1,4-polybutadiene ESR spectra with some resolved hyperfine structure are observed (Carstensen, 1971*a*). The spectra consist of a broad main signal (as shown in Fig. 1*a*) with a *g*

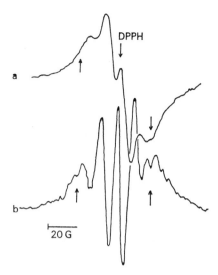

Fig. 1. ESR spectrum of *cis*-1,4-polybutadiene irradiated with UV light and recorded at 77 K immediately after irradiation. *a*, First derivative; *b*, second derivative. The short arrows indicate the positions of the lines which disappear on heating the sample. The center of the ESR signal of crystalline DPPH is marked as a long arrow.

value about free spin and two narrow signals with a separation of 510 G. The narrow lines are due to hydrogen atoms formed in, or adsorbed to, the surface of the quartz glass (Carstensen & Rånby, 1967). The radical concentration increases rapidly at the beginning of the irradiation and then levels off. By increasing radiation doses the spectra become asymmetric. The broad main signal (Fig. 1) consists apparently of a signal with an even number of lines of which four lines in the center are observed. By recording the second derivative of the ESR absorption more details are seen (Fig. 1*b*). The spectrum consists mainly of an even number of lines, presumably not bigger than six. This sextet spectrum with about 12 G hyperfine splitting is attributed to allyl free radicals (1) formed by chain scission midway between two double bonds. This radical is predominant in the polybutadiene samples irradiated at 77 K

$$—CH_2—CH{=}CH—\dot{C}H_2 \leftrightarrow —CH_2—\dot{C}H—CH{=}CH_2 \tag{1}$$

By a slight heating (114 K for 3 min) the two lines marked as arrows in Fig. 1 disappear. The lines, which have a separation of 50 G and a *g* value somewhat bigger than free spin, are probably a part of a spectral component with more than these two lines. The structures of the remaining lines are blurred by the strong sextet. The spectrum is attributed to an instable radical (2) formed by hydrogen abstraction from the strongly twisted main chain. The free p orbital will not be parallel with the π orbital system of the double bond and resonance stabilization will not occur. On heating to temperatures where segmental motions are possible, this radical will rearrange and resonance stabilize (3):

$$—CH_2—CH{=}CH—\dot{C}H— \tag{2}$$
$$\downarrow$$
$$—CH_2—CH{=}CH—\dot{C}H— \leftrightarrow —CH_2—\dot{C}H—CH{=}CH— \tag{3}$$

By heating to higher temperatures the resolution is improved without any change in radical concentration. By heating to temperatures just below the glass-transition temperature (T_g) the radical concentration decreases, and at temperatures above T_g the ESR signal has changed to a weak singlet. This signal is attributed to polyenyl radicals (4) containing a series of conjugated double bonds along the polybutadiene chains. These radicals are formed either by the direct action of light or during the warming towards T_g of the polybutadiene samples irradiated at 77 K.

$$—(CH_2—CH{=}CH)_n—CH_2^. \tag{4}$$

To investigate whether the presence of oxygen has any influence on radical formation, some irradiations were performed in the presence of air. The spectra were identical with those recorded after irradiation in vacuum, both in shape and intensity. After heating the samples to temperatures just below T_g the spectra changed and the characteristic peroxy radical absorptions were observed. The peroxy radicals cannot account for the asymmetry of the spectra recorded after long irradiation times. It is therefore not likely that oxygen plays any role in the formation of radicals by UV irradiation of polybutadiene.

Free Radicals in *cis*- and *trans*-1,4-Polyisoprene

By γ irradiation of natural rubber at 77 K, several authors (Ohnishi et al., 1960, 1961; Tsvetkov, Molin & Voevodskii, 1959) observed singlet-like ESR spectra without any hyperfine structure. When *trans*-1,4-polyisoprene was irradiated in vacuum at ambient temperatures, ESR spectra with some resolution were observed (Kuzminskii et al., 1962). However, when *cis*-1,4 polyisoprene was irradiated at the same conditions, no radicals were detected. Some differences in the ESR spectra of irradiated natural rubber and polyisoprene have been reported (Kozlov & Tarasova, 1966). The spectra in the above mentioned studies cannot unambiguously be related to any specific radical structure because of the lack of resolved hyperfine structure.

Only few ESR studies have been reported on the formation of free radicals in polyisoprene initiated by UV light (Milinchuk, 1965; Kimmer et al., 1963; Carstensen, 1970a). But as in the case of *cis*-1,4-polybutadiene hyperfine structures are observed in the ESR spectra after UV irradiation of *cis*- and *trans*-1,4-polyisoprene (Carstensen, 1970a). The spectra of the two *cis*- and the *trans*-samples were identical. The signals are similar in shape to those observed for *cis*-1,4-polybutadiene, but small differences are observed: (*a*) The intensity ratio between the central lines of the spectrum is bigger for polybutadiene than for polyisoprene; (*b*) the total width of the signal is bigger for polyisoprene than for poly butadiene.

Even in this case the second derivative spectra are valuable for the assignment of ESR spectra to radical structures (Carstensen, 1970 a). The spectrum consists of eight lines with a hyperfine separation of 12 G and is accordingly attributed to main chain scission radicals (5, 6):

$$-CH_2-\underset{\underset{CH_3}{|}}{C}=CH-CH_2^{\cdot} \tag{5A}$$

$$-CH_2-\underset{\underset{CH_3}{|}}{\overset{\uparrow\downarrow}{C}}-CH=CH_2 \tag{5B}$$

$$^{\cdot}CH_2-\underset{\underset{CH_3}{|}}{C}=CH-CH_2-\!\!-\!\!- \tag{6A}$$

$$CH_2=\underset{\underset{CH_3}{|}}{\overset{\uparrow\downarrow}{C}}-CH-CH_2- \tag{6B}$$

By heating the samples to temperatures just below T_g the resolution is increased without any noteworthy change in spectral shape. By heating to temperatures above T_g the spectra change to a singlet. This singlet is attributed to polyenyl radicals (7) formed either during irradiation or during warming up.

$$-(CR=CH)_n-CH_2^{\cdot} \tag{7}$$

where R is alternatingly H and CH_3, and $n=3$.

Free Radicals in *cis*-1,4-Polypiperylene and Polychloroprene

Polypiperylene was chosen for further work because of its structural analogy with both polyisoprene and polybutadiene. As commonly known a C–C bond is weaker than a corresponding C–H bond. This means that if hydrogen abstractions occur during UV irradiation of polybutadiene and polyisoprene there should be an even higher probability that methyl radicals are formed during irradiation of polypiperylene. The halflife of methyl radicals in a polymer matrix at 77 K, e.g. polypropylene (Rånby & Yoshida, 1966), polymethylmethacrylate (Kato & Nishioka, 1966), and polyisobutylene (Carstensen, 1967), is bigger than 1 h. This means that methyl radicals, if formed, will be observed in the ESR spectra.

The first derivative ESR spectrum (Carstensen, 1971b) has a smooth outline without any trace of hyperfine structure. It is therefore concluded that methyl radicals are not formed and accordingly that the probability of hydrogen abstraction in the case of polybutadiene and polyisoprene is extremely low.

The second derivative ESR spectrum is imperative to the assignment of the free radicals formed. The spectrum is apparently composed of an even number of lines (hyperfine splitting 12 G), but the intensity distribution implies the presence of another component with an uneven number of lines. The spectra are accordingly attributed (Carstensen, 1971*b*) to main chain scission radicals (8, 9)

$$CH_3$$
$$|$$
$$—CH_2—CH\!=\!\!CH—CH^{\cdot}$$

$$\uparrow\downarrow \qquad CH_3$$
$$|$$
$$—CH_2—\overset{\cdot}{C}H—CH\!=\!\!CH \qquad\qquad (8)$$

$$CH_3$$
$$|$$
$$^{\cdot}CH_2—CH\!=\!\!CH—CH—$$

$$\uparrow\downarrow \qquad CH_3$$
$$|$$
$$CH_2\!=\!\!CH—\overset{\cdot}{C}H—CH— \qquad\qquad (9)$$

The line width of the octet (radical (8)) is smaller than the line width of the quintet (radical (9)). This is probably due to a lower energy of activation for the rotation around the [–CH$_2$–CH–] bond in (8) than around the [–CH–CH(CH$_3$)–] bond in (9). Polychloroprene shows after irradiation with UV light a broad ESR spectrum and even in the second derivative spectrum only traces of hyperfine structure are observed. The samples were dark coloured after irradiation, but after annihilation of the radicals by warming to room temperature the samples were strongly yellow. This implies that polyenyl radicals are [formed] during irradiation and that the signal observed mainly is a singlet with a superimposed hyperfine structure from other radical species. The monomer units in polychloroprene are, like in other polymers used in this study, not in the same plane but merely ordered in a helix. This means that if the process of conjugation starts with a hydrogen abstraction from the chain a simultaneous movement of several neighbouring monomer units must occur, which is highly improbable at these low temperatures. On the other hand, if the process starts with a main chain scission, a movement of the new chain end is sufficient to continue the conjugation process.

Despite the lack of hyperfine structure in the ESR spectra it might be suggested that the primary radiation products are chain scission (radicals (10, 11)):

$$Cl$$
$$|$$
$$—CH_2—C\!=\!\!CH—CH_2^{\cdot}$$

$$Cl \;\uparrow\downarrow$$
$$|$$
$$—CH_2—\overset{\cdot}{C}—CH\!=\!\!CH_2 \qquad\qquad (10)$$

$$\overset{\displaystyle \overset{\textstyle Cl}{\underset{\textstyle |}{}}}{·CH_2—C=CH—CH_2—}$$

$$\overset{\displaystyle \overset{\textstyle Cl}{\underset{\textstyle |}{}} \updownarrow}{CH_2=C—\overset{\prime}{C}H—CH_2—} \tag{11}$$

These allyl radicals will have a UV absorption maximum around 2580 Å (Bodily & Dole, 1966a, b) and absorption of new UV quanta by these groups are therefore probably resulting in the formation of hydrogen gas and polyenyl radicals.

Mechanism of Radical Formation

There is good evidence that the free radicals in the polymers studied are formed by direct interaction of UV photons with the polymers. Repeated purification of the polymers did not change the rate of radical formation. The initial absorption of UV light may be due to the double bonds present (Carr & Walker, 1936) or to oxidized structures formed during polymerization or storage. The rate and efficiency of radical formation is of the same order of magnitude for all the polymers except for polychloroprene in which the rate of radical production was much larger. The effect of oxygen on the free radical formation during UV irradiation was studied in separate experiments. These experiments have given no evidence that oxygen takes part in the primary reaction of radical formation in these 1,4-polydienes.

It is noteworthy that all the samples showed no photochemistry above 3000 Å, in accordance with the findings of Golub & Stephens (1968) and Bateman (1945). This means that the active wavelength range in this study is from 2 400 to 3 000 Å, i.e. the photons absorbed raise the molecular energy by 120 to 95 kcal mol^{-1}, which is more than the dissociation energies of normal C–C and C–H bonds. The energy of each photon would not be sufficient for breakage or opening of more than one chemical bond for each event. The occurrence of chiefly main chain scission may be related to the expected preference of UV light to break the weakest bonds (Carstensen, 1970b). The bond strength of the C–C bonds connecting two successive monomer units is approx. 55 kcal mol^{-1} (Golub, 1971), which is lowered from the normal C–C bond by the resonance energy of the two allyl radicals formed by chain scission.

Iwasaki et al. (1967, 1969a, b) have reported that radicals formed by chain scission of polymers lead to an ESR transition at $g=4$ as expected for radical pairs. In the present study no signal at $g=4$ was observed because of the small amount of radicals formed by UV irradiation. The magnetic dipole/dipole interactions between the two unpaired electrons in the chain scission radicals will cause a broadening of the ESR lines. When the samples are heated, segmental motions will be possible and the radicals can either react or move away from each other resulting in a decrease in the dipole/dipole broadening. The irreversible line narrowing observed after heating the samples to temperatures below their respective T_g indicates that movements of the newly formed chain end radicals occur.

Degradation Mechanisms

The irradiated samples used for the ESR measurements in this study did not show any gas evolution or any new absorption lines or changes in absorption intensities of the IR spectra. However, others have found that when 1,4-polybutadiene and -isoprene were irradiated with 2 537 Å in vacuum at room temperature changes in microstructure occur. The most important changes are *cis-trans* isomerization (Golub & Stephens, 1968) and loss of original double bonds (Golub & Stephens, 1968). In addition, crosslinking (Golub, 1967; Golub & Stephens, 1967), formation of hydrogen gas (Bateman, 1946), vinyl (Golub & Stephens, 1967), and cyclopropyl groups (Golub & Stephens, 1968) were observed. As for 1,4-polyisoprene, only 1,1,2-trisubstituted cyclopropyl groups (Golub & Stephens, 1968) were formed together with new vinylidene double bonds (Golub & Stephens, 1967).

It is evident from the microstructural changes and the interpretation of the ESR spectra that there is a marked similarity between the radiation processes of the different 1,4-polydienes when exposed to UV light. As similar reaction schemes can be drawn up for both 1,4-polybutadiene and 1,4-polyisoprene explaining the products formed by irradiation, 1,4-polyisoprene is chosen to illustrate the degradation mechanisms.

It might seem contradictory that these polymers preferentially form chain scission radicals by irradiation as crosslinking is observed. However, Charlesby (1955) has shown that where active end groups are formed by chain fracture, a network structure will result. The energy absorbed by the polymer is used to rupture the weakest bond in the polymer chain, which is the one connecting two successive monomer units, resulting in the formation of two active chain end radicals (5) and (6). These radicals can (a) recombine in any of four different ways forming the original double bond in either *cis-* or *trans-* configuration or new vinyl and vinylidene groups or (b) add to a double bond forming an end-link under the loss of a double bond. As for the *cis-trans* isomerization of 1,4-polyisoprene, this reaction may be a consequence of the allyl resonance. The allyl radicals would have to possess an excess energy of approx. 17 kcal mol^{-1} (Golub & Gargiulo, 1972) in order to achieve a rotation about the single bonds in (5B) and (6B). As the energy of the incident light is approx. 100 kcal mol^{-1} and the strength of the C–C bond in question is approx.

55 kcal mol^{-1} it is possible that such "hot" radicals are formed in the photolysis of 1,4-polydienes.

$$
\begin{array}{ccc}
\underset{-CH_2}{\overset{CH_3}{\diagdown}}C=CH\underset{CH_2}{\diagup} & \leftrightarrow & \underset{-CH_3}{\overset{CH_3}{\diagdown}}\dot{C}-CH\underset{CH_2}{\diagup} & \rightarrow & \underset{-CH_2}{\overset{CH_3}{\diagdown}}\dot{C}-CH\overset{CH_2}{\diagup}
\end{array}
$$

(5 A)

$$
\leftrightarrow \quad \underset{-CH_2}{\overset{CH_3}{\diagdown}}C=CH\overset{CH_2}{\diagup}
$$

(5 B)

To explain the formation of cyclopropyl groups a biradical mechanism followed by a 1,2-hydrogen migration and a ring closure has been proposed (Golub & Stephens, 1968):

$$
-CH_2-\overset{\overset{\textstyle CH_3}{|}}{C}=CH-CH_2- \xrightarrow{h\nu} -CH_2-\overset{\overset{\textstyle CH_3}{|}}{\dot{C}}-\dot{C}H-CH_2- \rightarrow
$$

$$
-CH_2-\overset{\overset{\textstyle CH_3}{|}}{\dot{C}}-CH_2-\dot{C}H- \rightarrow -CH_2-\overset{\overset{\textstyle CH_3}{|}}{C}\underset{CH_2}{\diagdown}{\diagup}CH-
$$

As severe limitations must be put on the hydrogen migration to account for the formation of only 1,1,2-trisubstituted cyclopropyl groups, an alternative mechanism has been proposed (Carstensen, 1970*a*): The radicals (5, 6) formed by chain scission might undergo a direct ring closure with the formation of cyclopropyl radicals:

$$
-CH_2-\overset{\overset{\textstyle CH_3}{|}}{C}=CH-\dot{C}H_2 \longrightarrow -CH_2-\overset{\overset{\textstyle CH_3}{|}}{C}\underset{CH_2}{\diagdown}{\diagup}\dot{C}H \tag{5}
$$

$$
\dot{C}H_2-\overset{\overset{\textstyle CH_3}{|}}{C}=CH-CH_2- \longrightarrow \dot{C}\underset{CH_2}{\overset{\overset{\textstyle CH_3}{|}}{\diagdown}}{\diagup}CH-CH_2- \tag{6}
$$

The cyclopropyl radicals can saturate by recombination with e.g. their counterpart radicals. In this way only one type of cyclopropyl groups will be formed If so, the allyl radicals should possess an excess energy of 26 kcal mol^{-1} (Golub, 1971) in order to overcome the energy of the ring closure. As stated above such "hot" radicals might arise because of the large energy difference between the incident photons and the bond strength.

The formation of hydrogen gas is reported to have a quantum yield of 40 to 80 times smaller than the loss of original unsaturation. This is in accordance with the findings in this work as radicals formed by hydrogen abstraction were not observed. The low yields of hydrogen may be due to the formation of polyenyl radicals:

$$
\begin{array}{cc}
CH_3 & CH_3 \\
| & | \\
-CH_2-C=CH-CH_2-CH_2-C=CH-CH_2 \xrightarrow{h\nu} & \qquad (5)
\end{array}
$$

$$
\begin{array}{cc}
CH_3 & CH_3 \\
| & | \\
-CH_2-C=CH-CH=CH-C=CH-CH_2 + H_2 & \qquad (7)
\end{array}
$$

The allyl radicals (5, 6) will have a UV absorption maximum around 2 580 Å (Bodily & Dole, 1966a, b), which makes absorption of new photons probable.

Furthermore, it has been shown (Golub, 1967; Golub & Stephens, 1967) by use of deuterated samples that double bond shift does not occur; this means that abstraction of hydrogen atoms are not likely to occur as the radicals formed in this way are the only ones which could be responsible for the double bond shift. This study shows that the scission of the C–C bonds midway between two monomer units in the 1,4-polydienes is more important than any other process in producing free radicals when exposed to UV light. These chain scission radicals may account for all the microstructural changes observed in these polymers. Of course this does not exclude the possibility that some of the changes occur during spontaneous reactions via non-radical processes.

References

Bateman, L., Trans. Inst. Rubber Ind., *21*, 118 (1945).
Bateman, L., Trans. Faraday Soc., *42*, 267 (1946).
Bodily, D. M. & Dole, M., J. Chem. Phys., *44*, 2821 (1966 *a*).
Bodily, D. M. & Dole, M., J. Chem. Phys., *45*, 1428 (1966 *b*).
Carr, E. P. & Walker, M. K., J. Chem. Phys., *4*, 751 (1936).
Carstensen, P. & Rånby, B., Radiation Research 1966 (ed. G. Silini) p. 291. North Holland Publ., Amsterdam 1967.
Carstensen, P., Makromol. Chem., *135*, 219 (1970 *a*).
Carstensen, P., Acta Polytech. Scand., Chem. incl. Metall. Ser., No. 97, 1970*b*.
Carstensen, P., Makromol. Chem., *142*, 131 (1971 *a*).
Carstensen, P., Makromol. Chem., *142*, 145 (1971 *b*).
Charlesby, A., Proc. Roy. Soc. (London), *A231*, 251 (1955).
Golub, M. A., Radiation Research 1966 (ed. G. Silini) p. 339. North Holland, Amsterdam 1967.
Golub, M. A. & Stephens, C. L., J. Polymer Sci. C, *16*, 765 (1967).
Golub, M. A. & Stephens, C. L., J. Polymer Sci. A-1, *6*, 763 (1968).

Golub, M. A., Lecture given at the IUPAC Conference on Chemical Transformations of Polymers, Bratislava June 1971. To be published in Pure and Applied Chemistry.

Golub, M. A. & Gargiulo, R. J., J. Polymer Sci. B, *10*, 41 (1972).

Iwasaki, M. & Ichikawa, T., J. Chem. Phys., *46*, 2851 (1967).

Iwasaki, M., Ichikawa, T. & Ohmori, T., J. Chem. Phys., *50*, 1984, (1969 *a*).

Iwasaki, M., Ichikawa, T. & Ohmori, T., J. Chem. Phys., *50*, 1991 (1969 *b*).

Kato, Y. & Nishioka, A., Rep. Prog. Polymer Physics, Japan, *9*, 477 (1966).

Kimmer, W., Voelkel, G., Wartewig, S., & Windsch, W., Plaste Kautschuk, *10*, 345 (1963).

Kozlov, V. T. & Tarasova, Z. N., Vysokomol. Soedin., *8*, 943 (1966).

Kuzminskii, A. S., Neiman, M. B., Fedoseeva, T. S., Lebedev, Ya. S., Buchachenko, A. L. & Chertkova, V. F., Dokl. Akad. Nauk. SSSR, *146*, 611 (1962).

Kuzminskii, A. S., Fedoseeva, T. S. & Buchachenko, A. L., Kauchuk i Rezina, *7*, 10 (1965).

Libby, D., Ormerod, M. G. & Charlesby, A., Polymer (London), *1*, 212 (1960).

Milinchuk, V. K., Vysokomol. Soedin., *7*, 1293 (1965).

Ohnishi, S., Ikeda, Y., Sugimoto, S. & Nitta, I., J. Polymer Sci., *47*, 503 (1960).

Ohnishi, S., Ikeda, Y., Kashiwagi, M. & Nitta, I., Polymer (London), *2*, 119 (1961).

Rånby, B. & Yoshida, H., J. Polymer Sci. C, *12*, 263 (1966).

Tsvetkov, Yu. D., Molin, Yu. N. & Voevodskii, V. V., Vysokomol Soedin., *1*, 1805 (1959).

Discussion

Tsuji

You proposed that UV absorbers are double bonds and oxidative groups. I believe, however, that the double bonds are not responsible for UV absorption in your study since the absorption due to the double bonds starts at about 200 nm.

Carstensen

A smooth absorption tail extending to around 310 nm observed in the ultra-violet absorption spectrum of purified polyisoprene films (Carr & Walker, 1936) could involve direct excitation of the polyisoprene double bonds by the incident ultraviolet photons without intervention of inter- or intramolecular energy transfer processes.

D. R. Smith

In the early 1960's C P Poole reported ESR studies of radicals formed by UV photolysis at 77 K of low molecular weight olefins, linear and branched. In every case, I believe, he saw allyl or substituted allyl radicals formed by loss of H from the carbon adjacent to the double bond. I do not think he saw any C–C bond breakage. Can you comment on these observations in relation to your results?

Carstensen

In low molecular weight olefins and in polydienes, the single bonds include many with dissociation energies differing strongly from those in alkanes. The electron delocalization in the allyl radical fragments formed by bond rupture reduces the dissociation energy by about 20 kcal mol^{-1} for each allyl radical formed. This means that in the olefins studied by C P Poole the dissociation energy of the C–H and C–C bonds leading to allyl radicals will be reduced with the same amount. The radicals formed by C–C bond rupture are bulky and diffusion from the reaction site is not possible at these low temperatures; it is therefore likely that these radicals will recombine before detection and only H-abstraction radicals are observed.

As to the 1,4-polydienes, breakage of the C–C bonds midway between two double bonds is favoured, because here two resonance stabilized allyl radicals are formed, which means that the dissociation energy is approx. 55 kcal mol^{-1}. As for the corresponding C–H bond scission the dissociation energy is of the order 80–100 kcal mol^{-1}, because the polymer chains are frozen in a non planar conformation making electron delocalization of the allyl radical fragment impossible. Therefore radicals formed by C–C bond scissions are more likely to be observed in the polydienes than in the low molecular weight olefins.

Rabek

Generally, much amount of oxygen is absorbed by polybutadiene. Have you removed it?

Carstensen

The sample which we used is a commercial product. We have removed anti-oxidants, but we have not checked the oxygen absorbed.

Sohma

What is the mechanism of the temperature increase? Is it by irradiation or by heat transfer?

Carstensen

The apparent increase in signal intensity on heating the samples towards their respective T_g is due to a decrease in line width and not to a real increase in radical concentration.

Sohma

Is it oxygen which causes different effects by heating?

Carstensen

No, it is molecular movements. The increase by dipole–dipole interaction is not the intensity but the width of the peak.

ESR Studies of UV-Induced Degradation of Polyolefins

By Z. Joffe and B. Rånby

Department of Polymer Technology, The Royal Institute of Technology, Stockholm, Sweden

The degradation of polyolefins initiated by UV irradiation has been intensively studied in recent years using electron spin resonance (ESR) analysis (Yoshida & Rånby, 1964; Browning, Ackermann & Patton, 1966; Rånby & Yoshida, 1966; Campbell, 1970). In earlier work (Smaller & Matheson, 1968; Molin et al., 1958; Koritskii et al., 1959) high energy irradiation was used to induce radical reactions in polyolefins and other polymers, and the data obtained have facilitated the interpretation of the later UV irradiation studies. In the case of polyolefins, irradiation gives very different free radical reactions compared with high energy irradiation. High energy irradiation penetrates deep into the polymer samples and causes free radical formation both in the crystalline and the amorphous phase. The free radicals in the crystalline phase have generally rather long life time. Therefore, the concentration of free radicals is usually sufficiently high for studies by the ESR method, even after rather short irradiation times at moderate temperatures. UV irradiation, which is considerably less effective and less penetrating than high energy irradiation (Charlesby & Thomas, 1962), induces radical formations in polyolefins predominantly in amorphous and "defect" crystalline polymer phases, located in the surface layer of the polymer samples. The resulting radicals are usually very reactive and disappear easily by recombination. After UV irradiation at moderate temperatures, the momentary concentration of free radicals is usually too low to be detected by the ESR method. Therefore, it is necessary to accumulate the radicals formed, e.g. by trapping them at low temperature during the UV irradiation. By this means, the radical concentration can be sufficiently high for ESR investigation.

The presently accepted concepts for the UV initiated reaction of polyolefins are based on Norrish type I (free radical reaction) and type II (photochemical processes) (Trozzolo & Winslow, 1968; Calvert & Pitts, 1966). These photodegradation processes are related to carbonyl groups occurring randomly along the polymer chain, e.g. effects of polymerization or subsequent oxidation reactions. At moderate temperatures, the Norrish type II reaction seems to be dominant for polyolefins. At low temperatures, however, and especially below the glass transition temperature (T_g), the type I process becomes more important (Hartley & Guillet, 1968):

$$-CH_2-\overset{\overset{\displaystyle O}{\|}}{C}-CH_2- \xrightarrow{h\nu} -CH_2-\overset{\cdot}{C}=O + \cdot CH_2-CH_2-$$
$$\text{(III)} \qquad\qquad \text{(II)}$$

It is also known that photooxidation of polyolefins at low temperatures may initiate reactions which lead to competition between scission and crosslinking of the chains and depend on the presence of contaminations or modified groups, as mentioned previously.

Winslow et al. (1968) have suggested that the carbonyl groups in polyolefins may initiate further oxidation by formation of excited oxygen molecules, e.g. in a singlet state, which react with double bonds in the chains:

$$^1(>C=O) \xrightarrow{h\nu} {}^1(>C=O)^* \rightarrow {}^3(>C=O)^*$$

$$^3(C=O)^* + O_2 \rightarrow {}^1(>C=O) + {}^1O_2$$

$$^1O_2 + RCH_2CH=CH_2 \rightarrow RCH=CH-CH_2OOH$$

The hydroperoxides formed can decompose and yield more carbonyl groups, repeating the cycle to cause further degradation and oxidation (Bolland, 1946). However, after UV irradiation of polyolefins in high vacuum or at low O_2 pressure, it was very difficult to find oxidized radicals in the ESR spectra (Tsuji & Seiki, 1970; Rånby & Joffe, 1972). At high O_2 pressure, oxidized radicals were detected by ESR measurements (Tsuji & Seiki, 1971) and found to be unstable during warming up to moderate temperatures.

Polyethylene (PE)

Rånby & Yoshida (1966) and Browning et al. (1966) first reported results of ESR measurements for PE, using UV light as irradiation source. After UV irradiation of high density PE (HDPE) at $-196°C$ in vacuo, Rånby & Yoshida (1966) obtained a spectrum containing a diffuse sextet and assigned it to alkyl radicals formed by hydrogen abstraction (Fig. 1).

$$-CH_2-\overset{\cdot}{C}H-CH_2- \qquad\qquad\qquad\qquad\qquad \text{(I)}$$

with a hyperfine splitting constant $a_H = 33$ G. This result is in good agreement with earlier identifications of PE radicals, obtained by high energy irradiation (Smaller & Matheson, 1958; Molin et al., 1958; Tsvetkov et al., 1958; Bresler et al., 1959).

Furthermore, traces were found of a quintet spectrum with $a_H = 30$ G, as-

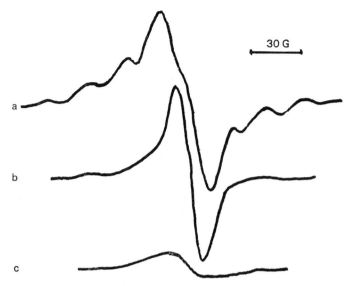

Fig. 1. ESR spectra of HDPE irradiated with UV light at −196°C (*a*) immediately after ir-radiation; (*b*) after warming to −95°C for 10 min; (*c*) after warming to 0°C for 10 min.

signed to radicals (II) supposed to be formed by chain breaks.

—CH_2—$\dot{C}H_2$

Both spectra (I) and (II) showed magnetic anisotropy.

ESR measurements of γ-irradiated PE, using the modified technique of "differential saturation" (Grinberg, Dubinskii & Lebedev, 1971), have later been presented. The spectrum obtained by this technique is a pure quintet with a separation of about 30 G (Fig. 2) and attributed to radicals of type (II). This confirms the results from UV irradiation previously reported by Rånby & Yoshida (1966).

Similar results have been obtained for UV irradiated low density (LD) PE (Rånby & Joffe, 1972) of thermally relaxed samples (Fig. 3).

Tsuji & Seiki (1971) also assumed the formation of type (II) radicals in ESR spectrum of PE at −196°C, but proposed a radical conformation with un-equivalence of the β hydrogens.

In the early work on radical formation in PE (Rånby & Yoshida, 1966) it was pointed out that the energy level of UV quanta (250–370 nm) is of the same order of magnitude as the strength of the common chemical bonds in polymers, e.g. C–H, C–C, C–O. Therefore, the breaking of C–C bonds by UV irradiation in the main chain is energetically possible. It was assumed that the absorption of UV quanta was due to oxidized groups, double bonds or impurities in the polymer. Later work has shown that the main reaction of breaking C–C bonds in the polymer chains, is indeed related to $>C=O$ groups in excited states.

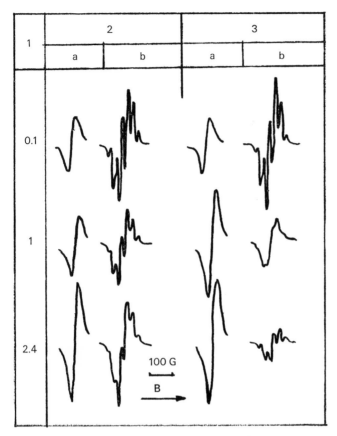

Fig. 2. "Sequence" and "differential" saturation of PE γ irradiated at $-196°C$. *1*, detector current [mA]; *2*, sequence saturation: (*a*) standard sample ($CaCl_2 \times 2$ H_2O); (*b*) signal; *3*, differential saturation: (*a*) standard sample (cf *2a*); (*b*) signal.

ESR spectra of acyl type radicals (III) obtained from γ irradiated acyl chloride in a glassy matrix at $-196°C$ as a strong singlet (or a triplet with line separation of 8 G) with a *g* value of 2.001, have been previously described by Noda, Fueki & Kuri (1968). Very similar singlets have been obtained using UV irradiated aliphatic ketones with long chains (symmetrical and unsymmetrical with C_{12} to C_{16}) in a glassy matrix of tetrahydrofurane (THF) or in solid state at $-196°C$ (Rånby & Joffe, 1972) (Figs 4, 5).

As interpretation of a study on UV irradiation of ketones (Hama & Shinohara, 1970, Hama et al., 1971) have proposed a mechanism for UV degradation of PE at low temperatures, with the acyl radical (III), undergoing a reversible conversion to some type of alkyl radical (IV)

$$-CH_2-\overset{\cdot}{C}H-CHO \underset{\text{light}}{\overset{\text{heat}}{\rightleftarrows}} -CH_2-\overset{\cdot}{C}=O$$

$$\text{(IV} \qquad\qquad \text{(III)}$$

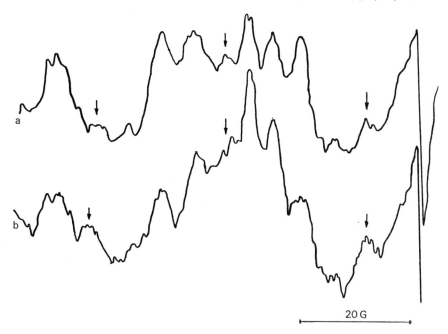

Fig. 3. ESR spectra of LDPE UV irradiated at $-196°C$ for 1 h (sample was thermally relaxed before irradiation). (*a*) immediately after UV irradiation; (*b*) after warming to $-159°C$ for 5 min.

The identification of the free radicals according to this mechanism, however, does not seem to be conclusive (Tsuji & Seiki, 1971; Rånby & Joffe, 1972). For ketones irradiated with UV light at $-196°C$ and warmed up to $-120°C$ in a glassy matrix, Rånby & Joffe (1972) have observed the appearance of a poorly resolved triplet (or quintet) (Fig. 6 A), present also after irradiation of the samples in the solid state (Fig. 6 B).

In their extensive ESR study of HD and LDPE irradiated with UV light, Tsuji & Seiki (1971) have reported the formation of the following radicals as a function of the conditions:

At $-196°C$ type (I), (II) and (III) radicals of which typ (III) radicals appear after longer irradiation decay

$$-CH_2-\overset{\cdot}{C}=O \xrightarrow{h\nu} -\overset{\cdot}{C}H_2 + CO$$
$$\quad\;\;(III) \qquad\qquad\quad\; (II)$$

At about $-130°C$ in air and then at $-78°C$ also in air, peroxy radicals,

$$-CH_2-\underset{\underset{\overset{|}{O_2^{\cdot}}}{}}{CH}-CH_2-$$

which are represented in ESR spectrum by a typical asymmetric spectrum with g value $g_{\parallel}=2.032$, $g_{\perp}=2.004$ (Fig. 7).

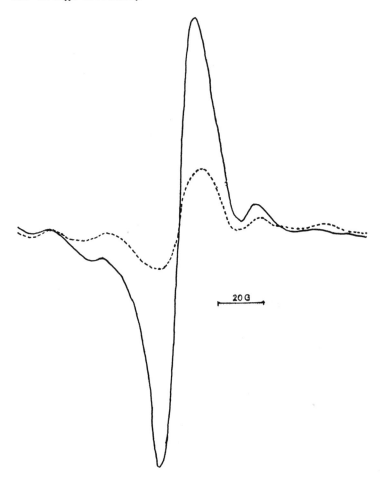

Fig. 4. ESR spectrum of dioctyl ketone $[CH_3(CH_2)_7]_2CO$ UV irradiated at $-196°C$ for 1 h, in glassy matrix of THF. *g* value 2.001. ----, THF at $-196°C$; ——, irradiation time 1 h.

At $-125°C$ (in vacuum) acyl radicals (VI) but with different structure:

$$-CH_2-\overset{\cdot}{C}H-CH_2- + CO \xrightarrow{h\nu} -CH_2-CH-CH_2-$$
$$\underset{\cdot CO}{|}$$

(I)   (VI)

type (VI) radicals were represented in the ESR spectrum by a strong singlet with a *g* value of 2.001.

At $-70°C$ and higher temperatures, type (VII) radicals, represented in ESR spectrum by quintet with a separation of about 12 G and a *g* value of 2.003:

$$-CH_2-\overset{\cdot}{C}H-CH=CH-CH_2-$$ (VII)

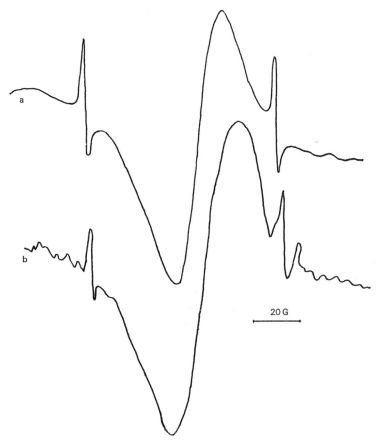

Fig. 5. ESR spectra of dioctyl ketone [CH$_3$(CH$_2$)$_7$]$_2$CO UV irradiated at $-196°$C in solid state. g value 2.0009 ± 0.0001 (*a*) 15 min; (*b*) 1 h after irradiation.

For UV irradiated PE, it has been reported (Tsuji & Seiki 1970) some influence of the O$_2$ pressure on the temperature of maximum intensity of the strong singlet spectrum at about T_g temperature. A considerable increase in radical concentration in irradiated PE, according to increase of the O$_2$ pressure from 10^{-5} to 10^{-1} torr at about T_g temperature have also been observed by Rånby & Joffe (1972) (Fig. 8).

Furthermore, the presence of oxidized PE radicals in UV irradiated LDPE using the ESR saturation method published by Ohnishi, Sugimota & Nitta (1963), have been recorded (Rånby & Joffe, 1972) in spite of low O$_2$ pressure on the samples.

Polypropylene (PP)

ESR studies of UV irradiated PP (Yoshida & Rånby, 1964; Browning, Ackermann & Patton, 1966; Rånby & Yoshida, 1966) show other types of radicals than there obtained with high energy irradiation (Campbell, 1970).

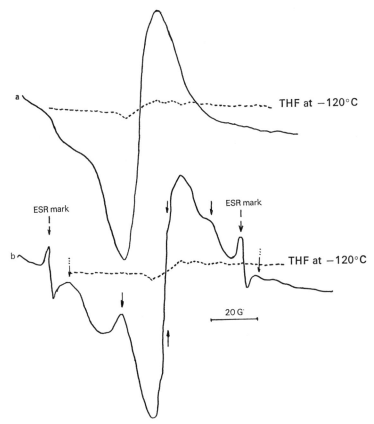

Fig. 6 A. ESR spectra of UV irradiated ketones at −196°C and warmed up to −120°C. Ketones in glassy matrix (THF):
(*a*) dioctyl ketone; (*b*) ethyl undecyl ketone.

Yoshida & Rånby (1964) and Rånby & Yoshida (1966) have examined with UV irradiation, isotactic PP at −196°C and pressure about 10^{-4} torr on the samples. The authors (Yoshida & Rånby, 1964; Rånby & Yoshida, 1966) have observed that the ESR spectrum of such irradiated PP sample consists of two components: a quartet with sharp lines and relatively short life time, interpreted as arising from methyl radicals (VIII) (Fig. 9)

$$\cdot CH_3 \qquad\qquad\qquad\qquad\qquad\qquad\qquad (VIII)$$

and another component of the spectrum, containing broad lines which was assigned to type (I) radicals

$$-CH_2-\dot{C}H-CH_2- \quad (I)$$

together with

$$-CH_2-CH-CH_2- \qquad\qquad -CH-\dot{C}H-CH-$$
$$\big| \qquad \text{and/or} \qquad \big| \qquad\qquad \big|$$
$$\cdot CH_2 \qquad\qquad\qquad\qquad CH_3 \qquad\quad CH_3$$
$$(IX) \qquad\qquad\qquad\qquad\qquad\qquad (X)$$

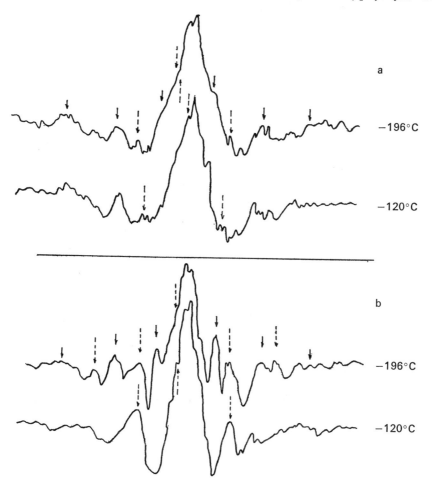

Fig. 6B. ESR spectra (2nd deriv.) of UV irradiated ketones at −196°C for 1 h and warmed up to −120°C:
a, dioctyl ketone; *b*, ethyl-undecyl ketone.

Rånby & Yoshida (1966) considered also that the broad line quartet could be due to radicals typ (XI)

$$—CH_2—CH—\dot{C}H_2$$
$$\underset{\displaystyle CH_3}{\vert} \qquad (XI)$$

formed by main chain scission, analogous with the interpretation for PE. The sharp component due to methyl radicals is represented by an equally spaced quartet line spectrum with separation of 22.5 G, and intensity ratio of 1:3:3:1, $\Delta H_{msl} = 3.4 \pm 0.1$ G, while the broad component shows a hf separation of 21 G, and $\Delta H_{msl} = 20$ G. Immediately after UV irradiation, the

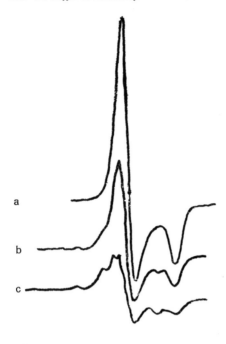

Fig. 7. Photo-induced conversion of peroxy radicals in HDPE, irradiated in oxygen atmosphere at $-196°C$. (*a*) Peroxy radicals were produced by treatment of the irradiated sample (*a*) at $-78°C$ for 1 min; (*b*) after UV irradiation (longer than 280 nm) for 9 min; (*c*) for 39 min.

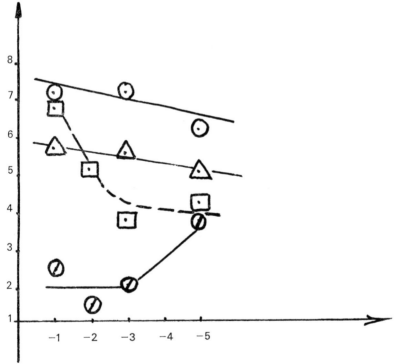

Fig. 8. *Abscissa:* log P_{O_2}; *ordinate:* $10^{-17} \times N_x$ [rad/cm³]. $\oslash-\oslash$, $-73°C$; $\square-\square$, $-126°C$; $\triangle-\triangle$, $-159°C$; $\bigcirc-\bigcirc$, $-196°C$.

Changes in radical concentration of UV irradiated LDPE at $-196°C$ and warmed up to $-73°C$ as a function of O_2 pressure on the samples.

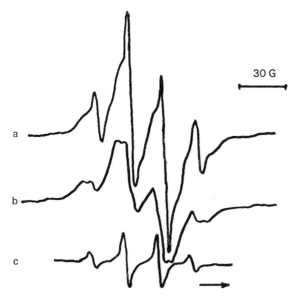

Fig. 9. ESR spectra of PP irradiated with UV light at $-196°C$. (*a*) immediately, (*b*) 2 days after irradiation; (*c*) the component which decayed during the first 2 h after irradiation.

methyl radicals constituted about 10 % of the average radical concentration, measured with the ESR spectrometer. It was also evident from the ESR measurement that PP, irradiated with UV at $-196°C$, gave about seven times more radicals than HDPE under similar conditions (Rånby & Yoshida, 1966). This phenomenon can be interpreted as due to the presence of a tertiary carbon in the PP chain, which is much more sensitive to hydrogen abstraction (Fitton, Howard & Williamson, 1970):

$$CH_3-\overset{\displaystyle -CH_2}{\underset{\displaystyle -CH_2}{>}}CH \xrightarrow[\text{or R.}]{h\nu} CH_3-\overset{\displaystyle -CH_2}{\underset{\displaystyle -CH_2}{>}}C^{\cdot} \quad \text{fast}$$

and

$$\overset{\displaystyle -CH_2}{\underset{\displaystyle -CH_2}{>}}CH_2 \xrightarrow[\text{or R}^{\cdot}]{h\nu} \overset{\displaystyle -CH_2}{\underset{\displaystyle -CH_2}{>}}\dot{C}H \quad \text{slow}$$

Browning et al. (1966) also photolyzed PP with UV light (253.7 nm) and observed similar spectra to those of Rånby & Yoshida (1966). The assignments of the observed radicals are in rather good agreement.

Some dependence of pressure for the amount of methyl radicals in PP after UV irradiation, has been described by Tsuji & Seiki (1970). They observed that PP samples UV irradiated at $-196°C$ and a pressure of $10^{-6}-10^{-5}$ torr,

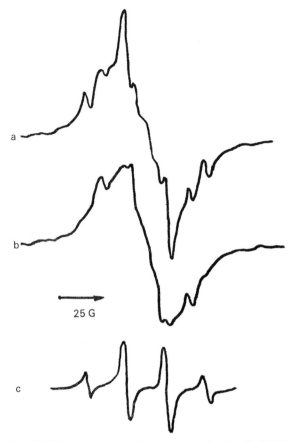

a

b

25 G

c

Fig. 10. ESR spectra of polyisobutene irradiated with UV light at −196°C (a) immediately after irradiation; (b) after keeping the sample for 5 days at −196°C; (c) the spectral component which decayed during the first hours at −196°C.

did not get a sharp component in the ESR spectrum due to methyl radicals. Only a quartet with coupling constant of about 25 G attributed mainly to free radicals of type (IX) and partially to radicals type (I) was observed. However, when the pressure in the sample tubes was 10^{-2} torr or higher, methyl radicals were present in the ESR spectrum. Tsuji & Seiki (1970) were of the opinion that the methyl radicals disappear quickly by UV irradiation in a highly evacuated PP sample.

Even if these radicals were produced as a primary event, they could be non-perceptible in the ESR spectra. This is, however, contradiction with earlier published work (Yoshida & Rånby, 1964; Browning, Ackermann & Patton, 1966; Rånby & Yoshida, 1966).

Other Polyolefins

ESR spectra of polyisobutene after UV irradiation at −196°C are different from those obtained after high energy irradiation (Rånby & Carstensen 1967;

Carstensen & Rånby, 1967) and were interpreted as due to radicals formed by hydrogen and methyl group abstraction, while no main-chain scission radicals were indicated. The following radicals have been recognized:

Main chain $-\dot{C}H-$ (XIII), side group $-\dot{C}H_2$, methyl radicals $\dot{C}H_3$ (in analogy with UV-irradiated PP) (Rånby & Yoshida, 1966), and their main-chain counterpart

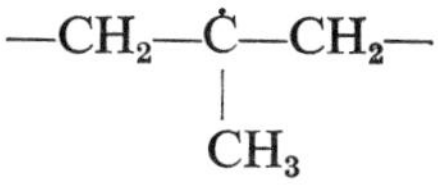

$$-CH_2-\overset{\displaystyle |}{\underset{\displaystyle CH_3}{\dot{C}}}-CH_2- \qquad (XIII)$$

In addition, a doublet with splitting of about 510 G was observed, and ascribed to hydrogen atoms formed in the sample or in the quartz glass sample tube and adsorbed on the glass surface. Poly(1-butene), irradiated with high energy irradiation at $-196°C$ gave a poorly resolved ESR spectrum with six lines. Its hf splitting constant of 21 G indicates that the spectrum represents the following radical structure (Loy, 1963):

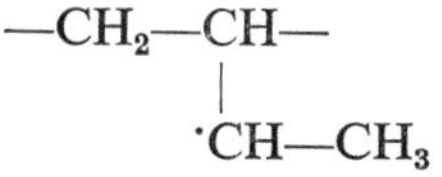

$$-CH_2-\overset{\displaystyle |}{\underset{\displaystyle \cdot CH-CH_3}{CH}}- \qquad (XIV)$$

but by using UV light as a source of irradiation, only ethyl radicals $\dot{C}H_2CH_3$ (XV) giving a sextet ESR spectrum with hf splitting constant of 27 G, have been reported (Ohnishi, Sugimota & Nitta, 1963).

High energy-irradiated poly(4-methyl-1-pentene) at $-196°C$ shows in ESR measurements a sextet with a hf splitting constant of 23 G (Kosumota, Hukuda & Takayanagi, 1965). As supposed, the spectrum is reported as due to radicals formed by side-chain scission:

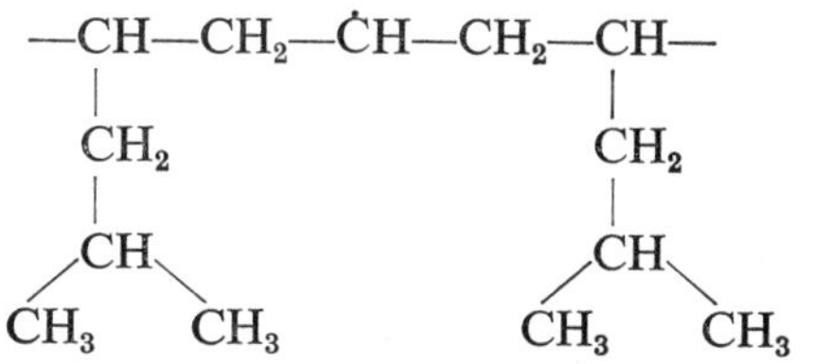

$$-\underset{\underset{\underset{\overset{\displaystyle |}{CH_3}\ \overset{\displaystyle \diagdown}{CH_3}}{\diagup CH\diagdown}}{\underset{\displaystyle |}{CH_2}}}{CH}-CH_2-\dot{C}H-CH_2-\underset{\underset{\underset{\overset{\displaystyle |}{CH_3}\ \overset{\displaystyle \diagdown}{CH_3}}{\diagup CH\diagdown}}{\underset{\displaystyle |}{CH_2}}}{CH}- \qquad (XVI)$$

The same polymer irradiated with UV light gives an ESR spectrum composed of a sharp quartet with a splitting constant of 22.5 G and a broad quartet (Browning, Ackerman & Patton, 1966). The sharp component has been attributed to methyl radicals (VIII), while the broad component could represent polymer radicals of the following structure:

$$-CH_2-CH-$$
$$\hspace{1.2em}|$$
$$\hspace{1.2em}CH_2$$
$$\hspace{1.2em}|\hspace{2em}\text{and/or}$$
$$\hspace{1.2em}CH-CH_3$$
$$\hspace{1.2em}|$$
$$\hspace{1.2em}{}^{\cdot}CH_2$$

(XVII)

$$-CH-\overset{\cdot}{C}H-CH-$$
$$\hspace{1em}|\hspace{2em}|$$
$$\hspace{1em}C_3H_9\hspace{1.2em}C_4H_9$$

(XVIII)

Similar radicals were proposed for this polymer irradiated with a super high-pressure mercury lamp at $-196°C$ (Goodhead, 1971).

These investigations are part of a research programme supported by The Swedish Board for Technical Development and The Swedish Polymer Research Foundation.

References

Bolland, J. L., Proc. Roy. Soc., A, 1946, *186*, 218.

Bresler, S. E. et al., Zh. Tekhn. Fiz., *29*, 358 (1959).

Browning, H. L., Ackermann, H. D. & Patton, H. W., J. Polymer Sci., *A-1*, *4*, 1433 (1966).

Calvert, J. G. & Pitts Jr., J. N., Photochemistry, ch. 5 (1966).

Campbell, D., Macromolecular Reviews, *4*, (1970).

Carstensen, P. & Rånby, B., Radiation Research, 1966, North-Holland, Amsterdam 1967.

Charlesby, A. & Thomas, D. K., Proc. Roy. Soc. (London), *A 265*, 104 (1962).

Fitton, S. L., Haward, R. N. & Williamson, G. R., British Polymer J., *2*, No 5. (1970).

Goodhead, D. T., J. Polymer Sci., *A-2*, *9*, 999 (1971).

Grinberg, O. Ya., Dubinskii, A. A. & Lebedev, Ya. S., Dokl. Akad. Nauk SSSR, *1971*, 196 (3) 627–9.

Hama, Y. & Shinohara, K., Molecular Phys., *18*, No. 2, 279 (1970).

Hama, Y., et al., J. Polymer Sci., *A-1*, *9*, 1411 (1971).

Hartley, G. H. & Guillet, J. E., Macromolecules, *1*, 165 (1968).

Koritskii, A. T., et al., Vysokomol. Soedin., *1*, 1182 (1959).

Kosumota, N., Hukuda, K. & Takayanagi, M., Rep. Progr. Polymer Phys. Japan, *8*, 315 (1965).

Loy, B. R., J. Polymer Sci., *A-1*, 2251 (1963).

Molin, I. N., et al., Dokl. Akad. Nauk SSSR, *123*, 882 (1958).

Noda, S., Fueki, K. & Kuri, Z., J. Chem. Phys., *49*, No. 7, 3287 (1968).

Ohnishi, S., Sugimoto, S. & Nitta, I., J. Polymer Sci., *A-1*, 605 (1963).

Rånby, B. & Yoshida, H., J. Polymer Sci., *C*, 263 (1966).

Rånby, B. & Carstensen, P., Irradiation of Polymers, Repr. from Advances in Chemistry Series No 66 (1967).

Rånby, B. & Joffe, Z., unpublished work (1972).

Smaller, B. & Matheson, M. S., J. Chem. Phys., *28*, 1169 (1958).

Trozzolo, A. M. & Winslow, F. H., Macromolecules, *1*, 98 (1968).

Tsuji, K. & Seiki, T., Rep. Progr. Polym. Phys. Japan, XII, 507 (1970).

Tsuji, K. & Seiki, T., Polymer J. (Jap.), *1*, No. 1, 133 (1970).

Tsuji, K. & Seiki, T., Polymer J. (Jap.), *2*, 606 (1971).

Tsvetkov, J. D. et al., Dokl. Akad. Nauk SSSR, *122*, 1053 (1958).

Winslow, F. H. et al., Polymer Reprints, 9/1, 377, 1968, San Francisco.

Yoshida, H. & Rånby, B., Polymer Letters, *2*, 1155 (1964).

Discussion

Tsuji

(1) Pressure dependence of methyl radical formation in PP was found to be caused by thermal effect during irradiation.

(2) Recently we observed the wavelength dependence of radical formation in PP, and methyl radicals were found to be produced by the secondary reactions of the free radicals produced by the decomposition of oxidative groups (J. Polymer Sci. B *10*, 139–144 (1972)).

Joffe

From the paper presented in my review regarding this problem (Tsuji & Seiki, Polymer J (Japan) *1*, No. 133 (1970)) it was not clearly explained which parameter in their experiments got the most important dependence on the $\cdot CH_3$ radical behaviour in PP after UV irradiation. Furthermore, the results were in some contradiction to earlier published work about PP UV irradiation (Yoshida & Rånby, 1964; Browning, Ackermann & Patton, 1966; Rånby & Yoshida, 1966).

P. Smith

As far as I am aware there have been no studies of the course of the UV-induced degradation of a polymeric organic solid in which the effects of the presence of a mechanical strain in the UV-irradiated solid has been investigated. Of course, it is clear from the results of investigations similar to those of Dr Peterlin (p. 235) that one source of such effects could be the radicals which might be formed as a result of the mechanical strain. However, it seems that there might be other possible sources of effects, e.g. the dependence on mechanical strain of the physical properties of the solid as a reaction medium because of the important role diffusion can play in solid-state reactions. Do you know of any work which has involved the study of the UV-induced degradation of polymeric organic solids with the use of different mechanical conditions?

Joffe

No.

Tsuji

$H_3C\cdot$ is produced due to heating effect. $H_3C\cdot$ is produced with shorter waves. Perhaps this is the result of a second reaction. Alkoxy radical is produced with longer waves.

ESR Study of Polyethylene Irradiated with Ultraviolet Light and Electron Beams

By Kozo Tsuji and Seizo Okamura

Central Research Laboratory, Sumitomo Chemical Co., Ltd. Takatsuki, Osaka, and Department of Polymer Chemistry, Faculty of Engineering, Kyoto University, Kyoto, Japan

Introduction

In order to study a photodegradation mechanism of polymers, it is important to follow the behavior of free radicals as intermediate species. ESR spectroscopy is a powerful method for detecting and identifying free radicals. Several studies have been reported on ESR spectra of polymers irradiated with UV light (Charlesby & Thomas, 1962; Michel, Chapman &| Mao, 1966; Browning, Ackerman & Patton, 1966; Rånby & Yoshida, 1966; Rånby & Carstensen, 1967, Hughes & Coard, 1969). The formation mechanism of the free radicals, however, have not been elucidated for all polymers studied, in spite of its importance. We are now carrying out studies on photodegradation mechanism of polymers, especially the primary processes of radical formation, by using the ESR technique. Results on polyethylene (Tsuji & Seiki, 1969, 1971 a, b), polypropylene (Tsuji & Seiki, 1970 a, b, 1972 a), poly-3,3-bis(chloromethyl)-oxetane (Tsuji, Hayashi & Okamura, 1970; Tsuji & Seiki, 1972 b) and poly-(2,6-dimethylphenylene oxide) (Tsuji & Seiki, 1972 c) have partly been reported.

It is also interesting to compare the behavior of the free radicals produced by UV irradiation and ionizing radiation in connection with the primary processes of radical formation, since UV light and ionizing radiation are considerably different (Tsuji 1972 a). In this paper we want to report the behavior of free radicals produced in polyethylene after irradiation with UV light and electron beams.

Results and Discussions

Free Radicals Produced in Polyethylene Irradiated with UV Light

(a) *ESR Spectrum Observed at* $-196°C$: Nitrogen-filled samples of both low-(LDPE) and high-density polyethylene (HDPE) gave eight-line spectra as shown in Figs 1 a and 2 a after irradiation in the liquid nitrogen Dewar at $-196°C$. This spectrum has been attributed to free radicals $-CH_2-\dot{C}H_2$

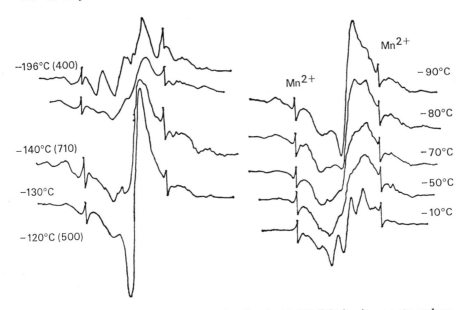

Fig. 1. Change of ESR spectrum of LDPE irradiated with UV light in nitrogen atmosphere with increasing temperature: irradiation time 2 h; irradiation temperature −196°C. Number in brackets represent amplifier gain settings. The separation between the two Mn²⁺ peaks is 86.7 G.

(I) and $-CH_2-\dot{C}H-CH_2-$ (II) (Tsuji & Seiki, 1971 *a*). Very recently, however, we obtained the evidence to support that this spectrum is due to single radicals of the type $-CH_2-\dot{C}H-CH_3$ (III) (Tsuji, 1972 *b*) rather than the superposition of the two free radicals (I) and (II). This result will be described later in connection with photo-induced radical conversions.

The wavelength dependence of radical formation was studied, and it was found that clear ESR spectra of irradiated polyethylene began to appear when the sample was irradiated with UV light of a wavelength longer than about 280 nm.

The same eight-line spectra were observed from both samples in N_2 and in O_2 after irradiation at −196°C. It is remarkable, however, that the intensity of the ESR spectrum of the sample in O_2 was greater than that in N_2 for all wavelengths studied here. This might be attributed to the extra absorption of light due to the charge-transfer complexes between polyethylene and oxygen molecules (Tsuji & Seiki, 1971*a*, *b*, 1970*b*). The difference of the absorption spectra of polyethylene films in O_2 and in N_2 supports this possibility.

It was found that the rate of radical formation is linear to the light intensity, when the light intensity is weaker.

(*b*) *Spectral change on warming:* Fig. 1 shows change of the ESR spectrum observed when the low-density polyethylene sample irradiated in N_2 was warmed successively to $-10°C$. At $-140°C$ the eight-line spectrum became poorly resolved and at $-130°C$ drastic change started to give a prominent singlet spectrum at $-120°C$ whose g value was 2.001. This singlet spectrum gradually decayed out at a higher temperature. At $-10°C$ there remained an apparent five-line spectrum with a separation of about 12 G. Upon further warming this spectrum was proved to be a seven-line spectrum with a substructure of separation of about 4 G and the g value was 2.003. This spectrum can reasonably be attributed to allylic radicals $-CH_2-\dot{C}H-CH=CH-CH_2-$ (IV). This spectrum, however, became a broad singlet when measured at $-196°C$.

The sharp singlet spectrum observed at about $-120°C$ was also observed after treatment of the irradiated sample at $-78°C$ for a short time. This spectrum was initially identified as polyenyl radicals (Rånby & Yoshida, 1966). Observed instability and g value (2.001), however, are not expected for radicals of this kind (Tsuji & Seiki, 1970 c). The g value is very close to the average of g values of several kinds of acyl radicals (Tsuji & Seiki, 1971 a). Therefore it seems correct that the sharp singlet spectrum observed here is due to some kind of acyl radicals. The probable structure of the acyl radicals is

$$\dot{C}=O \qquad\qquad (V)$$
$$|$$
$$-CH_2-CH-CH_2-$$

This identification was confirmed by the fact that the similar sharp singlet spectrum was also observed at about $-120°C$ during an analogous warm-up process of the sample irradiated with electron beams in the presence of CO at $-196°C$ (Tsuji & Takeshita, 1972). This result indicates that acyl radicals were produced by addition of CO to alkyl radicals.

The same warm-up experiment was carried out for the HDPE sample, and the spectral change was almost similar to that for LDPE as shown in Fig. 2. But the spectra at lower temperatures were better resolved and this made the assignment of the spectra easier. A six-line spectrum was observed at about $-135°C$ and $-125°C$ with a separation of about 30 G. This spectrum is similar to that observed after irradiation of polyethylene with ionizing radiation at $-196°C$ and attributed to the alkyl radicals (II). One noticeable point is that the intensity of the singlet spectrum observed during the warm-up process for HDPE was smaller than that for LDPE. This may reflect the fact that an amorphous region is smaller in HDPE than in LDPE, where CO molecules can diffuse easily.

On the other hand, the sample irradiated in air or in O_2 gave an asymmetric spectrum characteristic of peroxy radicals when the sample was warmed to $-130°C$, although the same eight-line spectrum was observed at $-196°C$ as the sample irradiated in N_2 (Fig. 3). The g values of the spectrum due to the

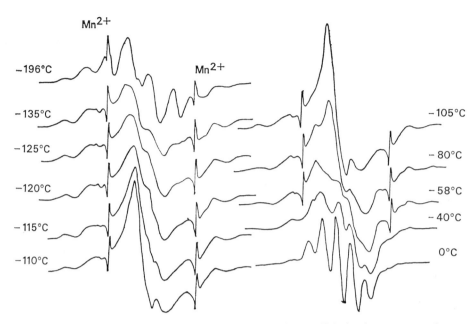

Fig. 2. Change of ESR spectrum of HDPE irradiated with UV light in nitrogen atmosphere with increasing temperature: irradiation time 90 min; irradiation temperature $-196°C$. The separation between the two Mn^{2+} peaks is 86.7 G.

peroxy radicals were $g_{||}=2.032$ and $g_{\perp}=2.004$. The peroxy radicals were also observed after treatment at $-78°C$ for a short time of the sample irradiated in the presence of air. Upon successive warming, the intensity of the peroxy radicals decreased gradually, and there remained an apparent five-line spectrum due to the allylic radicals at $0°C$.

(c) *Primary Processes of Radical Formation and Secondary Reactions of Free Radicals:* According to the fact that acyl radicals were produced at about $-120°C$ during the warm-up process it is apparent that CO molecules evolved after UV irradiation of polyethylene at $-196°C$. Therefore it is readily supposed that the primary process of radical formation in polyethylene is a Norrish type I reaction of carbonyl groups contained in the polymer. In fact the carbonyl groups were detected by infrared spectrum of a polyethylene film. It is also reported (Heskins & Guillet, 1970) that photochemical reactions of copolymers of ethylene and carbon monoxides as model compounds of polyethylene are those occurring in simple dialkyl ketones. Wavelength dependence of radical formation also supports this argument.

$$-CH_2CH_2CH_2CCH_2CH_2-\xrightarrow{h\nu}(CO+2-CH_2CH_2CH_2^*)\rightarrow-CH_2\dot{C}HCH_3$$
$$\overset{||}{O}\qquad\qquad\qquad\text{(I)}\qquad\qquad\qquad\text{(III)}$$

$$(1)$$

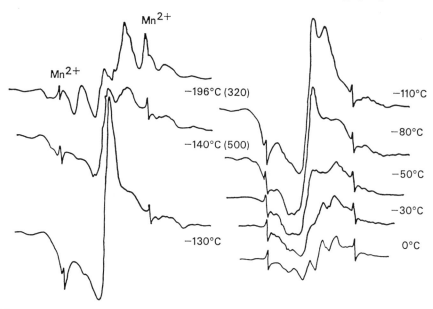

Fig. 3. Change of ESR spectrum of LDPE irradiated with UV light in air with increasing temperature: irradiation time 90 min; irradiation temperature −196°C. Numbers in brackets represent amplifier gain settings. The separation between the two Mn²⁺ peaks is 86.7 G.

The asterisk indicates some excited state of free radicals. The conversion of free radicals (I)* to (III) could be supposed easily when one takes into consideration that the free radicals (I)* might be produced as intermediate species in the reaction (4) described later (Shimada, Kashiwabara & Sohma, 1970).

At about −140°C the free radicals (III) changed into the free radicals (II). This conversion was also reported to take place in γ-irradiated high-density polyethylene, although at higher temperature (Shimada, Hashiwabara & Sohma, 1970). The addition of CO to the free radicals (II) took place at about −120°C, which corresponds to T_g of polyethylene, probably due to diffusion of CO molecules.

$$-CH_2\dot{C}HCH_3 \xrightarrow{\sim -140°C} -CH_2\dot{C}HCH_2- \qquad (2)$$

$$-CH_2\dot{C}HCH_2- + CO \rightarrow -CH_2\overset{\overset{\displaystyle \dot{C}=O}{|}}{C}HCH_2- \qquad (3)$$

At higher temperature allylic radicals were observed. This is caused by repeated hydrogen atom abstraction reactions by free radicals, which resulted in the formation of only the stable allylic radicals.

(d) *Effect of Aromatic Additives:* It is reported that some aromatic compounds were detected in polyethylene as impurities although in very small amounts

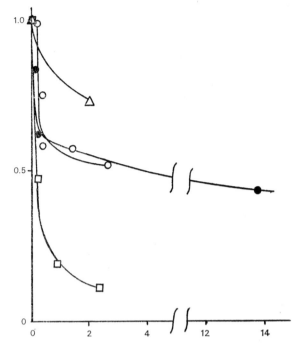

Fig. 4. *Abscissa:* additive conc. ($\times 10^3$ mol l^{-1}); *ordinate:* relative radical conc. $-\triangle-$, phenanthrene; $-\bullet-$, anthracene; $-\bigcirc-$, naphthalene; $-\square-$, pyrene.
Dependence of radical concentration after 20 min irradiation on the additive concentration. Irradiations were carried out directly from a Xe short arc lamp (no filter).

(Boustead & Charlesby, 1967). As some kinds of aromatic compounds are known to sensitize the photodecomposition of hydrocarbons, ethers, alcohols and polydimethylsiloxane (Judeikis & Siegel, 1965), it is important to examine the effect of aromatic impurities on the photo-induced radical formation in polyethylene.

When polyethylene samples containing aromatic compounds such as naphthalene, anthracene, phenanthrene or pyrene were irradiated with UV light applying no filter, the similar eight-line spectrum with somewhat poorer resolution was observed. The signal intensity, however, was found to decrease with additive concentration, and the sharp decrease was observed at additive concentration of an order of 3×10^{-4} mol l^{-1} (Fig. 4). These results could be explained reasonably by an optical filtering effect due to absorption by the aromatic molecules and the triplet-triplet energy transfer from the excited carbonyl groups to the aromatic molecules (Tsuji, Seiki & Takeshita, 1972). This result would also support the suggestion that the carbonyl groups are responsible for radical formation.

On the other hand, when irradiations were carried out with the light of wavelengths longer than about 300 nm, radical formation was sensitized by

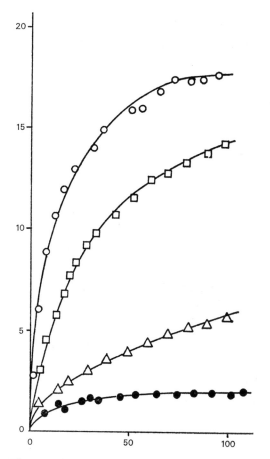

Fig. 5. *Abscissa:* irradiation time (min); *ordinate:* relative radical conc. $-\bigcirc-$, no filter; $-\square-$, phenanthrene UV-31; $-\triangle-$, quinoxaline UV-31; $-\bullet-$, UV-31 filter (> 300 nm).
Dependence of radical concentration on irradiation time for LDPE containing aromatic molecules (10^{-3} mol 1^{-1}).

addition of the aromatic compounds as shown in Fig. 5 (Takeshita, Tsuji & Seiki, 1972). In this case, however, the spectrum observed was a broad singlet with some shoulders, which is tentatively attributed to mainly the allylic radicals according to its g value (2.003). The shoulders are due to the alkyl radicals. The rate of radical formation was found to be approximately proportional to the square of the incident light intensity, which is a marked contrast to the linear dependence observed for radical formation in pure polyethylene. ESR spectra of the $\Delta M = \pm 2$ transition of the photo-excited triplet state of the aromatic molecules were also observed. Since the incident light is absorbed only by the aromatic molecules, these results suggest that the radical formation is related to a biphotonic absorption of the incident photons by the aromatic molecules, followed by energy transfer mainly to the $C{=}C$ bonds in polyethylene, which causes allylic radical formation. The excited triplet states of the aromatic molecules have the life time long enough to be intermediate species for biphotonic absorption of the incident light.

Free Radicals Produced in Polyethylene Irradiated with Electron Beams

(a) *ESR Spectrum Observed at* −196°C: When polyethylene was irradiated with electron beams at −196°C, a well-known six-line spectrum with the separation of about 30 G was observed. This spectrum is attributed to alkyl radicals (II). These results should be compared with those obtained for UV irradiation. In the case of ionizing radiation, the primary event of radiation is an ionization and/or excitation of molecules. ESR spectra due to trapped electrons (Keyser, Tsuji & Williams, 1968) and cation radicals (Campbell, 1970) in polyethylene have been reported. Several kinds of mechanisms of alkyl radical formation were suggested such as ion-molecular reactions or dissociation from the excited state (Timm & Willard, 1969).

(b) *Spectral Change on Warming:* As the sample irradiated with ionizing radiation in N_2 at −196°C was successively warmed, the signal intensity gradually decreased without any essential change in a spectral shape up to about −10°C. Above −10°C, however, the well-resolved seven-line spectrum of the allylic radicals appeared. This spectrum became a little ill-resolved when measured at −196°C. This is different from the result of UV irradiation. The sample irradiated at −196°C in the presence of air did not give the asymmetric spectrum due to peroxy radicals during the warm-up process or after treatment at −78°C for a short time. It is a marked contrast to the result obtained from polyethylene irradiated with ultraviolet light that neither the sharp singlet spectrum due to acyl radicals nor the asymmetric spectrum due to peroxy radicals were observed even after the same treatment as in the case of the sample irradiated with UV light.

When polyethylene was irradiated with electron beams at −196°C in the presence of CO, the same six-line spectrum was observed at −196°C as in the presence of N_2. This sample was subsequently warmed to give the sharp singlet spectrum at about −120°C due to the acyl radicals produced by addition of CO to the alkyl radicals (Fig. 6) (Tsuji & Takeshita, 1972). All alkyl radicals, however, did not convert into acyl radicals. This change is analogous to that observed after the same treatment of polyethylene irradiated with UV light at −196°C in the absence of CO as described before.

(c) *Trapping Regions of Free Radicals:* Free radicals produced by ionizing radiation are supposed to be trapped in the various regions, as ionizing radiation passes through the solid. From the sensitivity to oxygen, free radicals pro-

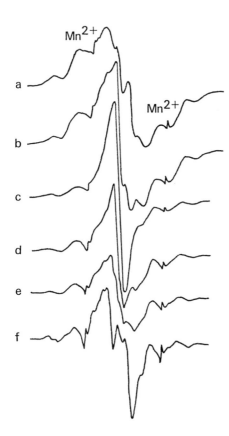

Fig. 6. Change of ESR spectrum of polyethylene irradiated at $-196°C$ in the presence of CO with increasing temperature. The spectra were measured at a, $-140°C$ (220); b, $-125°C$ (220); c, $-110°C$ (100); d, $-40°C$ (200); e, $-30°C$ (200); f, $0°C$ (500). Numbers in brackets represent amplifier gain settings. The separation between the two Mn^{2+} peaks is 86.7 G.

duced by UV irradiation and ionizing radiation are supposed to be trapped in different regions. In the case of UV irradiation, it is reasonable to suppose that the free radicals are trapped in the amorphous regions, where irregular groups, such as the carbonyl groups, are present and oxygen molecules can diffuse quite easily. On the other hand, free radicals produced after irradiation with electron beams seem to be trapped mainly in the crystalline regions where oxygen molecules cannot diffuse. It is interesting to note here that the trapped electrons produced after γ irradiation are trapped in the crystalline region (Keyser, Tsuji & Williams, 1968) in connection with the primary processes of radical formation by ionizing radiation. The observation that some of the free radicals reacted with CO might give more information on the trapping regions of free radicals.

Different temperature dependence of the ESR spectra due to allylic radicals produced by UV irradiation and ionizing radiation could be explained by the difference of the trapping regions of the free radicals described above.

(d) *Photo-Induced Radical Conversions:* It is reported that the allylic radicals produced after irradiation of HDPE with ionizing radiation at room tempera-

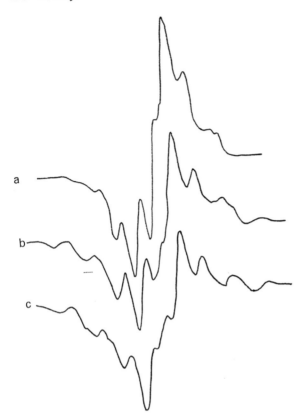

Fig. 7. Photo-induced radical conversion in polyethylene irradiated with electron beams under vacuum. *a*, spectrum due to allylic radicals; *b*, UV-irradiation, UV-39,218 min; *c*, UV-irradiation, no filter, 81 min.

ture convert to the alkyl radicals (II) by UV irradiation from low-pressure mercury lamps (Ohnishi, Sugimoto & Nitta, 1963). When a UV-39 filter was applied, on the other hand, conversion was reported to occur to a different type of alkyl radicals (III), which showed an eight-line spectrum (Shimada, Kashiwabara & Sohma, 1970). A similar experiment was carried out for LDPE (Tsuji, 1972). In this case the eight-line spectrum was observed after photolysis of the allylic radicals both with and without the filter (Fig. 7). This conversion was explained by the following reaction (Shimada, Kashiwabara & Sohma, 1970)

$$-CH_2-CH=CH-\dot{C}H-CH_2-CH_2-CH_2- \xrightarrow{h\nu}$$
$$-CH_2-CH=CH-CH=CH_2 + CH_3-\dot{C}H-CH_2- \quad (4)$$

It should be mentioned here that the eight-line spectrum observed after radical conversion is almost identical with that observed after UV irradiation of polyethylene at $-196°C$. Therefore the eight-line spectrum observed after photolysis of polyethylene at $-196°C$ could be attributed to the free radicals (III) as described before.

Photo-induced radical conversions of acyl radicals and peroxy radicals have already been reported (Tsuji & Seiki, 1971 *a*).

The authors are indebted to the company for permission to publish the results, and to Dr T. Takeshita and Mr T. Seiki for their collaboration in a part of this work.

References

Charlesby, A. & Thomas, D. K., Proc. Roy. Soc., *A269*, 104 (1962).
Michel, R. E., Chapman, F. W. & Mao, T. J., J. Chem. Phys., *45*, 4604 (1966).
Browning, H. L., Ackerman, H. D. & Patton, H. W., J. Polymer Sci., A-1, *4*, 1433 (1966).
Rånby, B. & Yoshida, H., J. Polymer Sci., C, No. 12, 263 (1966).
Rånby, B. & Carstensen, P., Advan. Chem. Ser., *66*, 256 (1967).
Hughes, O. R. & Coard, L. C., J. Polymer Sci., A-1, *7*, 1861 (1969).
Tsuji, K. & Seiki, T., J. Polymer Sci., B, *7*, 839 (1969).
Tsuji, K. & Seiki, T., Polymer, J., *2*, 606 (1971 *a*).
Tsuji, K. & Seiki, T., J. Polymer Sci., A-1, *9*, 3063 (1971 *b*).
Tsuji, K. & Seiki, T., Polymer, J., *1*, 133 (1970 *a*).
Tsuki, K. & Seiki, T., J. Polymer Sci., B, *8*, 817 (1970 *b*).
Tsuji, K. & Seiki, T., J. Polymer Sci., B, *10*, 139 (1972 *a*).
Tsuji, K., Hayashi, K. & Okamura, S., J. Polymer Sci., A-1, *8*, 583 (1970).
Tsuji, K. & Seiki, T., J. Polymer Sci., A-1, *10*, 123 (1972 *b*).
Tsuji, K. & Seiki, T., Rept. Prog. Polymer Phys. Japan, *15*, 573 (1972).
Tsuji, K. & Seiki, T., Rept. Prog. Polymer Phys. Japan, *15*, 559 (1972).
Tsuji, K., Rept. Prog. Polymer Phys. Japan, *15*, 555 (1972 *b*).
Tsuji, K. & Seiki, T., Rept. Prog. Polymer Phys. Japan, *13*, 507 (1970 *c*).
Tsuji, K. & Takeshita, T., J. Polymer Sci., B, *10*, 185 (1972).
Heskins, M. & Guillet, J. E., Macromolecules, *3*, 224 (1970).
Shimada, S., Kashiwabara, H. & Sohma, J., J. Polymer Sci., A-2, *8*, 1291 (1970).
Boustead, I. & Charlesby, A., Europ. Polym. J., *3*, 459 (1967).
Judeikis, H. S. & Siegel, S., J. Chem. Phys., *43*, 343, 3625 (1965).
Tsuji, K., Seiki, T. & Takeshita, T. J. Polymer Sci., A-1, *10* (1972). In press.
Takeshita, T., Tsuji, K. & Seiki, T., J. Polymer Sci., A-1, *10*, 2315 (1972).
Keyser, R. M., Tsuji, K. & Williams, F., Macromolecules, *1*, 289 (1968).
Campbell, D., J. Polymer Sci., B *8*, 313 (1970).
Timm, D. & Willard, J. E., J. Phys. Chem., *73*, 2403 (1969).
Ohnishi, S., Sugimoto, S. & Nitta, I., J. Chem. Phys., *39*, 2647 (1963).

Discussion

D. R. Smith

You mention that UV photolysis of an allylic radical caused it to be replaced by an alkyl radical. To support your interpretation I should mention a result

I published in Can. J. Chem. 4 or 5 years ago. I was able to form the substituted allylic radical $\overline{CH_2-C(CH_3)CHCH_2CH_3}$ from 2-methyl pentene-1, trapped in a 3-methylpentane matrix at 77 K. UV (2 537 Å) photolysis of this allylic radical induced abstraction to form the solvent radical $CH_3CH_2CH(CH_3)\dot{C}HCH_3$, exactly the same phenomenon as you have reported.

Tsuji

In our case, we suppose that the bond scission mechanism (eq. (4)) is probable as Professor Sohma et al. proposed (Shimada, Kashiwabara & Sohma, 1970).

Hedvig

The difference between the action of the ionizing radiation and UV illumination might be alternatively interpreted as being due to the different rates of the reaction

$$
\underset{\substack{\displaystyle | \\ \displaystyle O \\ \displaystyle | \\ }}{\overset{\displaystyle \dot{O}}{-C-C-C-}} \xrightarrow[\text{or } e^-]{h\nu} -\overset{|}{C}-\overset{|}{\dot{C}}\cdot \xrightarrow[\text{or } e^-]{h\nu} \overset{|}{C}-\overset{|}{\dot{C}}-\overset{|}{C}
$$

In polyethylene the chain-end radicals formed this way cannot be observed because the rate of the hydrogen abstraction reaction in the radiation field is very high. The same is true in the case of the carbonyl groups in your eq. (1).

I would not expect too much difference in the concentration of the carbonyl groups in the amorphous and crystalline phases, as most of these groups are formed during the course of preparation of the sample in the molten state before crystallization.

Tsuji

In the case of polyethylene also photolysis of peroxy radicals were observed to form alkyl radicals. We suppose the mechanism as follows

$$
\underset{\substack{\displaystyle | \\ \displaystyle H}}{\overset{\substack{\displaystyle \dot{O} \\ \displaystyle | \\ \displaystyle O \\ \displaystyle |}}{-CH_2-C-CH_2^-}} \xrightarrow{h\nu} -CH_2-\dot{C}H-CH_3
$$

which is very similar to that you suggested. The observed radicals, however, seem to be different between PTFE and PE.

We suppose that the carbonyl groups responsible for radical formation are mainly present in the amorphous regions because the free radicals produced with UV irradiation convert easily to peroxy radicals at about $-120°C$ while those produced after irradiation with electron beams do not. In the latter case free radicals seem to be trapped mainly in the crystalline regions to which oxygen molecules cannot diffuse.

P. Smith

Your suggestion that charge-transfer complexes between polyethylene and O_2 molecules may contribute to the photochemistry of polyethylene in an oxygen atmosphere seems very reasonable. Such complexes are even more likely to be formed by aromatic compounds than by aliphatic compounds. Do you know of any evidence that the contribution of such complexes may be significant in the photochemistry of polyethylene containing aromatic impurities? Of course, I realize that this question relates to a possible variation in what may be a small effect, but this variation might be comparatively large percentage wise.

Tsuji

We don't know of any evidence that the contribution of the CT complexes between aromatic impurities and oxygen molecules may be significant in the photochemistry of polyethylene. We suppose that the amount of such complexes is very small.

Rånby

Could you get polyenyl radicals?

Tsuji

No. We measured just after warming up in this study. Perhaps, after some time, an allylic radical may change to a polyenyl radical. In fact, when the sample showing the allylic radical spectrum was maintained at room temperature for 25 days, most of the ESR spectrum due to the allylic radicals disappeared and a broad singlet spectrum was obtained, which can be attributed to polyenyl radicals.

Photochemical Oxidation Reactions of Synthetic Polymers

By J. F. Rabek and B. Rånby

Department of Polymer Technology, The Royal Institute of Technology, Stockholm, Sweden

During the last few years, new mechanisms involving free-radical intermediates have been proposed for the photochemical oxidation of synthetic polymers, in bulk and in solution. Some of the main problems are related to initiation, propagation and termination reactions. The rate of oxidation depends also on the nature of the polymer, e.g. the crystalline and amorphous phase, branching, etc. Other factors influencing the process of polymer photooxidation are: contents of impurities, catalysts, photosensitizers, properties of solvents, the presence of singlet oxygen and ozone.

The Initiation Reaction

The formation of a polymer radical $P^{\cdot}$:

$$PH \rightarrow P^{\cdot} + H \tag{1}$$

is the necessary condition for a rapidly proceeding reaction of the polymer (PH) oxidation. The initiation reaction may be induced by

- (a) Direct dissociation of particular bonds in polymer molecule by ultraviolet irradiation;
- (b) Photochemical reaction of the carbonyl groups;
- (c) Molecular oxygen or oxygen-polymer charge-transfer complexes;
- (d) Singlet oxygen formed in the photochemical reactions.

Direct Dissociation of Bonds in Polymer by Ultraviolet Irradiation

The majority of common polymers have a low absorption of light, i.e. they are transparent to visible and UV light or they scatter light by physical heterogeneities. Saturated polyolefins in pure form are not expected to show UV absorption (Partridge, 1966; 1968), consequently they should be unaffected by sunlight in the atmosphere (wavelength >290 nm).

In the early work of Rånby et al. (Rånby & Yoshida, 1966; Yoshida & Rånby, 1964), the free radical formation due to scission of covalent chemical bonds caused by UV irradiation was studied as a primary effect, considering that the UV-light quanta have enough energy to break these bonds whatever was the light-absorption mechanism. The most common chemical bonds in polymers, C–H and C–C, have dissociation energies of 100 kcal mol^{-1} (4.3 eV) and 83 kcal mol^{-1} (3.6 eV) respectively, corresponding to UV-light quanta of the wavelengths 285 and 340 nm. The dissociation of a molecule occurs when a sufficient amount of energy accumulates in one bond. The threshold energy for the breaking of a bond is determined by the sum of the activation energy for dissociation and bond energy. The activation energy is closely related to the bonding force. In the papers mentioned above (Rånby & Yoshida) paid special attention to the investigation of free radicals which were produced by irradiation of polyethylene and polypropylene using ESR spectroscopy. The nature of the free radical obtained in these studies has been discussed (Joffe & Rånby, 1972).

Photochemical Reaction of the Carbonyl Groups

Because pure polyolefins show no absorption bands in the UV region it was suggested by Rånby et al. (Rånby & Yoshida, 1966; Yoshida & Rånby, 1964) and Tsuji & Seiki (1970*a*; 1971*b*; 1971*e*; 1971*f*) that the chemical effects of UV irradiation on these polymers were due to carbonyl groups or double bonds located along the polymer molecule. These groups may be formed by various uncontrolled reactions during the synthesis or the processing of the polymer. The introduction of carbonyl groups and systems of conjugated double bonds in the polymer results in a shift of the absorption bands towards longer wavelengths. In addition, carbonyl groups are easily excited in the range 270–330 nm to singlet and triplet states initiating a number of various photochemical reactions. The photochemical reactions of an excited carbonyl group in an organic compounds are classified as Norrish reactions of type I and type II. In the case of polymers these reactions cause degradation of the macromolecules.

Type I

The primary process, in which the bond between the carbonyl carbon and an adjacent α carbon is homolytically cleaved, is commonly called the α cleavage:

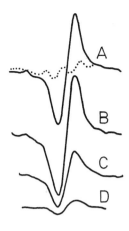

Fig. 1. ESR spectra of polyethylene irradiated by UV light. *A,* observed immediately at 77 K after irradiation; *B,* observed at 77 K after heat treatment of *A* at 163 K; *C,* observed at 77 K after heat treatment of *A* at 243 K; *D,* observed at 77 K after heat treatment of *A* at 303 K (Hama et al., 1971).

$$
\begin{array}{c}
\overset{R}{\underset{|}{-CH-CH_2-\overset{O}{\overset{\|}{C}}-CH_2-}} \xrightarrow{+h\nu}
\end{array}
$$

$$
\xrightarrow{+PH}\ \overset{R}{\underset{|}{CH-CH_2-C}}\overset{O}{\underset{H}{\diagup}} \qquad (5)
$$

$$
\rightarrow \overset{R}{\underset{|}{-CH-CH_2-CO^\cdot}} + {}^\cdot CH_2- \qquad (6)
$$
$$
(6a)
$$

$$
\searrow \overset{R}{\underset{|}{-CH-CH_2^\cdot}} + CO + {}^\cdot CH_2- \qquad (7)
$$

The type I photolysis of a ketone thus generates free radicals and also causes polymer chain scission. Acyl radicals (3a, 6a) are produced by UV irradiation from carbonyl groups located either at the chain ends or at some point along the chain. Last year Hama et al. (1971) studied the free acyl radicals (I) produced in UV-irradiated polyethylene using ESR measurements. By absorption of visible light these radicals were converted to alkyl radicals (II):

$$
-CH_2-CH_2-CO^\cdot \underset{heat}{\overset{+h\nu}{\rightleftarrows}} -CH_2-\overset{\cdot}{C}H-CHO \qquad (8)
$$
$$
(I) \qquad\qquad\qquad (II)
$$

The ESR spectra of polyethylene irradiated by UV light at 77°K are shown in Fig. 1. Noda et al. (1968) have reported the decomposition of low molecular acyl radicals to alkyl radicals and carbon monoxide. The same reaction can be written for a polymer acyl radical:

$$
-CH_2-CH_2-CO^\cdot \rightarrow -CH_2-CH_2^\cdot + CO \qquad (9)
$$
$$
(I) \qquad\qquad (III)
$$

The end group radical (III) can abstract a hydrogen atom from other polymer molecules even at $-196°$ C (Tsuji & Seiki, 1971 *c*):

$$-CH_2-CH_2^{\cdot}+-CH_2-CH_2-CH_2- \rightarrow$$
$$\text{(III)}$$

$$-CH_2-CH_3+-CH_2-\overset{\cdot}{C}H-CH_2- \quad (10)$$
$$\text{(IV)}$$

The ESR spectra of the radicals (III) and (IV) are observed after UV irradiation of polyethylene. The acyl radicals (V) observed during the warming of an irradiated sample are also formed in the reaction of alkyl radical (IV) with carbon monoxide (Tsuji & Seiki, 1971 *b*; 1971 *c*)

$$
-CH_2-\overset{\cdot}{C}H-CH_2-+CO \rightarrow -CH_2-\overset{\displaystyle \overset{\cdot CO}{|}}{C}H-CH_2- \quad (11)
$$
$$\quad\text{(IV)} \qquad\qquad\qquad\qquad \text{(V)}$$

Type II

Intramolecelar abstraction of γ hydrogen results in a decomposition generating an olefin and a ketone:

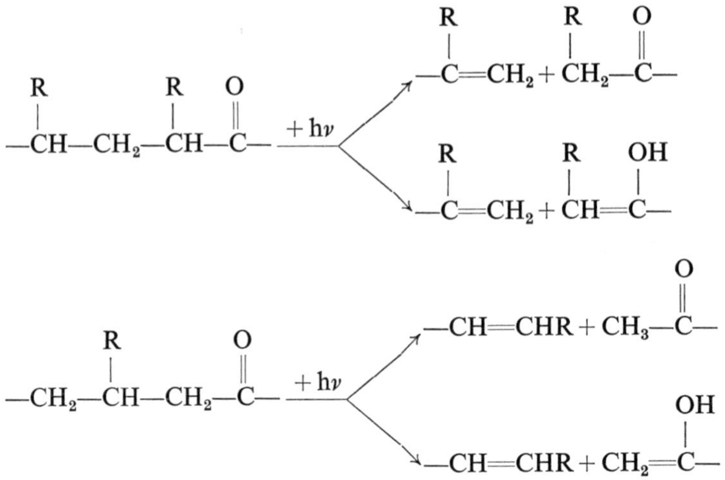

$$
\overset{\displaystyle R \quad\quad R \quad O}{\underset{}{-CH-CH_2-CH-C-}} \xrightarrow{+h\nu}
\begin{cases}
-\overset{\displaystyle R}{C}=CH_2+\overset{\displaystyle R}{C}H_2-\overset{\displaystyle O}{C}- & (12) \\[2ex]
-\overset{\displaystyle R}{C}=CH_2+\overset{\displaystyle R}{C}H=\overset{\displaystyle OH}{C}- & (13)
\end{cases}
$$

$$
\overset{\displaystyle R \quad\quad O}{\underset{}{-CH_2-CH-CH_2-C-}} \xrightarrow{+h\nu}
\begin{cases}
-CH=CHR+CH_3-\overset{\displaystyle O}{C}- & (14) \\[2ex]
-CH=CHR+CH_2=\overset{\displaystyle OH}{C}- & (15)
\end{cases}
$$

Norrish type I and type II reactions have been found a number of polymers such as polyketones (Amerik & Guillet, 1971; Guillet et al., 1968; Guillet & Norrish, 1955; Hartley & Guillet, 1968 *a*; Lukac et al., 1971), oxidized polyolefines (Carlsson et al., 1966; 1968; Ershov, 1969; Ershov et al., 1969; Heacock et al., 1968; Tsuji & Seiki, 1970 *a*), ethylene–carbon monoxide copolymers (Guillet et al., 1968; Hartley & Guillet, 1968 *a*; Heskins & Guillet, 1968; 1970), poly(*t*-butylacrylate) (Monahan, 1966) and poly(ethylene terphthalate) (Day & Wiles, 1972).

The quantum yields of the two types of Norrish reactions are related to the structure of ketone polymer:

(*a*) When the ketone group is located along a polypropylene chain, type I of the Norrish reaction predominates (Carlsson & Wiles, 1969 *a*):

$$CO + {}^{\cdot}CH_2-\underset{\underset{H}{|}}{\overset{\overset{CH_3}{|}}{C}}- \qquad (16) \quad \text{Type I A}$$

$$-\underset{\underset{H}{|}}{\overset{\overset{CH_3}{|}}{C}}-CH_2^{\cdot} + {}^{\cdot}\overset{\overset{O}{\|}}{C}-CH_2-\underset{\underset{H}{|}}{\overset{\overset{CH_3}{|}}{C}}- \qquad (17) \quad \text{Type II A}$$

$$-\underset{\underset{H}{|}}{\overset{\overset{CH_3}{|}}{CH}}-\overset{\overset{O}{\|}}{C}-CH_2-\overset{\overset{CH_3}{|}}{C}-CH_2-\underset{\underset{H}{|}}{\overset{}{C}}- \xrightarrow{+h\nu}$$

90% (to 16, 17)

10%

$$-CH=C\overset{\diagup CH_3}{\diagdown H}$$

$$-CH_2-C\overset{\diagup CH_2}{\diagdown H} \qquad + CH_2=\overset{\overset{OH}{|}}{C}-CH_2-\underset{\underset{H}{|}}{\overset{\overset{CH_3}{|}}{C}}- \qquad (18)$$

(*b*) When the ketone groups are located at the end of a polypropylene chain, the formation of acetaldehyde predominates in the reaction of Type I and of di-methylketone in the reaction type IIB (Carlsson et al., 1968; Carlsson & Wiles, 1969 *a*):

$${}^{\cdot}CH_3 + CO \qquad (19)$$

$$-CH_2-\overset{\overset{CH_3}{|}}{\underset{\underset{H}{|}}{C}}-CH_2- + {}^{\cdot}\overset{\overset{CH_3}{|}}{C}=O \rightarrow CH_3CHO \qquad (20) \quad \text{Type I B}$$

$$-CH_2-\overset{\overset{CH_3}{|}}{\underset{\underset{H}{|}}{C}}-CH_2-\overset{\overset{CH_3}{|}}{C}=O \xrightarrow{+h\nu}$$

12%

75%

$$-CH=C\overset{\diagup CH_3}{\diagdown H}$$

$$-CH_2-C\overset{\diagup CH_2}{\diagdown H} \qquad + (CH_3)_2C=O \qquad (21) \quad \text{Type II B}$$

The quantum yield of polypropylene photolysis in Norrish reaction of type IA is $\phi_{IA} = 0.07$ and that of Norrish reaction of type IB is $\phi_{IB} = 0.013$ (Carlsson & Wiles, 1969 *a*). These ϕ values are considerably higher than in the case of the photolysis of ethylene carbon monoxide copolymers with $\phi_I = 0.003$ (Hartley & Guillet, 1968*a*). The dissociation energy of the C–CO bond of the primary, secondary and tertiary ketones has been estimated to be $\pm 80 \pm 2$ kcal mol^{-1} (Benson, 1965). It is further assumed that C–CO in polymers is of the same magnitude.

On account of the high internal viscosity of the polymer the combination of the two macroradicals formed in reaction of type IA probably predominates over the RO_2 formation, especially when the concentration of the dissolved oxygen is low (cage effect). Radical combination according to the reaction type IB is unlikely, since $CH_3\dot{C}O$ and $\cdot CH_3$ radicals may diffuse from the reaction site before recapture. Other isolated macroradicals may also be formed by hydrogen abstraction by $\cdot CH_3$ and $CH_3\dot{C}O$ radicals.

The quantum yield of polypropylene photolysis in the Norrish reaction type IIA is $\phi_{IIA} = 0.01$ and that in Norrish reaction type IIB is 0.08 (Carlsson & Wiles, 1969 a), whereas the quantum yield of the photolysis of ethylene carbon monoxide copolymer in Norrish reaction type II is $\phi_{II} = 0.025$ (Hartley & Guillet, 1968 a). The reaction type II is assumed to proceed through a six-membered cyclic intermediate which is more readily formed by a terminal ketone than with long chain ketones of type A.

The type IIB dissociation of ketone from excited singlet or triplet states results in the liberation of a small molecule and generate unsaturation in the residue. Any dissociation of type II of a ketone results in the scission of the chain, though further photolysis of the ketone B formed does not contribute to the degradation of the chain.

Molecular Oxygen or Oxygen-Polymer Charge-Transfer Complexes

A direct interaction of the polymer with molecular oxygen following the abstracting of a hydrogen atom from a polymer molecule is however improbable because this is an endothermal reaction requiring an amount of energy of 30–40 kcal mol^{-1}. Recent proposals made by Tsuji & Seiki (1970 b; 1971 b, c, d, f) of the formation of charge transfer complexes between oxygen and a number of polymers, e.g. polyethylene, polypropylene and 3,3-*bis*(chloromethyl)oxetane may explain the bathochromic shift of the absorption range. Such complexes may absorb light in the range up to 340 nm and may form excited states. The absorbed energy may further induce the dissociation of the bond in the polymer. This problem is not yet fully elucidated and the role of charge-transfer complexes in the photooxidation processes awaits further investigations. There are other alternative theories which have been proposed for the initiation of the photooxidation of organic compounds and polymers.

Singlet Oxygen in the Photooxidation Reactions

During the last four years the role of singlet oxygen has received increasing attention in the photooxidation of polymers. The energy of excited singlet oxygen molecules in respect to the ground state of triplet oxygen can be estimated as follows: $^1O_2(^1\Delta_g)$ 22.5 kcal mol^{-1} (0.977eV), $^1O_2(^1\Sigma_g^+)$ 37.5 kcal mol^{-1} (1.63 eV). It was found by a number of investigators that various organic compounds such as aldehydes (Kummler & Bortner, 1969), ketones (Trozzolo, 1968), cyclic hydrocarbons (Algar & Stevens, 1970 a, b; Corey & Taylor,

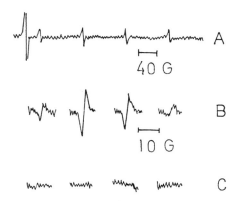

Fig. 2. ESR spectra of $^1O_2(^1\Delta_g)$, A, Singlet oxygen generated by a microwave discharge unit. The quartet due to $^1O_2(^1\Delta_g)$ is centered at $|g| \approx 2/3$. The intense line to the left is due to $^1O_2$ ($^1\Sigma_g^+$) oxygen; B, singlet oxygen $^1O_2(^1\Delta_g)$ generated by energy transfer from photoexcited naphthalene vapor; C, ESR signals observed under same conditions as B, except without irradiation of the cell (Kearns et al., 1969).

1964; Cowell & Pitts, 1968; Gleason et al., 1970; Star et al., 1969; Stevens & Algar, 1968; 1969) and dyes (Gleason et al., 1970; Gollnick, 1968 *a*, *b*; Kearns et al., 1967 *a*, *b*; Kearns, 1971; Stauff & Fuhr, 1969; Wilson, 1955) sensitize the photooxidation of many organic compounds. It is assumed that the singlet oxygen which takes part in the reaction formed by a reaction between the sensitizer molecule, excited to the singlet or to the triplet state and molecular oxygen. The chemistry of singlet oxygen is thoroughly discussed in several reviews (Gollnick, 1968 *a*, *b*; Rabek, 1971 *a*).

The ESR spectrum of $^1O_2(^1\Delta_g)$ was first observed in the products of a microwave discharge of O_2 by Fallick et al. (1965), who assigned the individual lines to the appropriate ($\Delta M_J = 1$) transition. The ESR spectrum in the $J = \Lambda = 2$ state is characterized by a symmetrical quartet of lines centered at $(\overline{g_J}) \sim 2.3$. Kearns et al. (1969) and Wasserman et al. (1968) observed the formation of ESR spectra of $^1O_2(^1\Delta_g)$ on UV irradiation of vapour-phase mixtures of oxygen with naphthalene and some of its derivatives. These ESR spectra are presented in Fig. 2.

The concept of singlet oxidation in the photooxidation of polymers was at first applied by Trozzolo & Winslow (1968). The following mechanism for these reactions was proposed:

1st stage: The UV radiation is absorbed by carbonyl groups causing their excitation (n, π^*).

2nd stage: Some of excited carbonyl group may initiate the cleavage (Norrish type II reaction) of carbon–carbon bonds and the formation of vinyl groups resulting in the degradation of the macromolecules:

$$-CH_2-\overset{\overset{\textstyle O}{\|}}{C}-CH_2-CH_2-CH_2-CH_2- \xrightarrow{+h\nu} \text{ketone } {}^3(n,\pi^*) \rightarrow$$

$$-CH_2-\overset{\overset{\textstyle O}{\|}}{C}-CH_3 + CH_2\!\!=\!\!CH-CH_2- \qquad (22)$$

3rd stage: Some of the excited carbonyl groups may transfer their excess energy to molecular oxygen forming the singlet oxygen in ${}^1O_2({}^1\Delta_g)$ state (Trozzolo, 1968):

$$\text{ketone } {}^3(n,\pi^*) + {}^3O_2 \rightarrow \text{ketone} + {}^1O_2 \qquad (23)$$

4th stage. Singlet oxygen may react with vinyl groups and form hydroperoxides:

$$CH_2\!\!=\!\!CH-CH_2-CH_2- + {}^1O_2 \rightarrow \overset{\overset{\textstyle OOH}{|}}{CH_2}-CH\!\!=\!\!CH-CH_2- \qquad (24)$$

Recent work (Kaplan & Kelleher, 1971) has demonstrated that treatment of polyethylene film with singlet oxygen in gas phase can produce the formation of hydroperoxides on the surface of the polymer. Kaplan & Kelleher (1970) have published a study of the oxidation of *cis*-polybutadiene by singlet oxygen. It has been suggested that singlet oxygen may be involved in the photooxidation of polyvinyl chloride (Kwei, 1969) and cellulose (Bourdon & Schuringer, 1967).

The photooxidation of polymers can be also induced by external photoinitiators and photosensitizers. In the first case molecules of photoinitiators are excited by absorbed light and subsequently decomposed into free radicals which may react with polymer molecules and abstract hydrogen atoms. In the case of photosensitizers which are excited by absorbed light, their excitation energy can be transferred to oxygen molecules forming singlet oxygen.

A series of investigations by Rabek have shown that several groups of compounds can sensitize the photooxidation of polydienes, e.g. diketones (Rabek, 1967b), aromatic ketones (Rabek, 1967a), quinones (Rabek, 1967c), peroxides (Rabek, 1967d), dyes (Rabek, 1971c; Rabek & Skorupa, 1971) and others. Mönig (1962; Mönig & Kriegel, 1960), Siewert (1964) and Fox & Priece (1967) found that addition of aromatic hydrocarbons (anthracene, pyrene, 3,4-benzopyrene) to polymethyl methacrylate in solution initiates the sensitized photooxidation of the polymer. Rabek (1968 b) has found that anthracene sensitize the degradation and photooxidation of polystyrene in solution. Recent experiments made in this laboratory have shown that singlet oxygen is responsible for the oxidative degradation in the presence of chloranile (Rabek & Rånby, 1972). The problem of sensitizing photooxidation of polymers has been extensively reviewed by Rabek (1968 a; 1971 c; 1972).

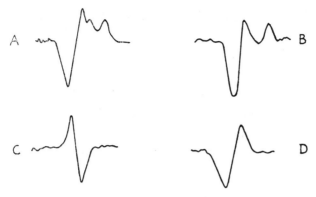

Fig. 3. ESR spectra of polypropylene peroxy radicals; *A*, at 110 °C; *B*, at 120°C; *C*, at 130°C; *D*, at 140°C. (Chien & Boss, 1967*a*).

The Propagation Reactions with Oxygen

Important problems in the photooxidation of polymers are related to the propagation reaction in the presence of oxygen. A mechanism for the fundamental oxidation process of polymer was proposed by Bolland (1949) and worked out in detail by Norling & Tobolsky (1970) and Tryon & Wall (1962).

Macroradicals formed during initiation react with oxygen molecules producing peroxy polymer radicals (PO_2·) according the reaction:

$$P· + O_2 \rightarrow PO_2· \tag{25}$$

Peroxy radicals of polymers have been detected by ESR, e.g. from polypropylene (Chien & Boss, 1967*a*, *b*) and from polytetrafluoroethylene (Iwasaki & Sakai, 1968; Moriuchi et al., 1970; Siegel & Hedgpeth, 1967). The ESR spectra observed are broad and symmetric (Fig. 3).

These spectra are characterized by three principal components of the anisotropic g tensor (g_1, g_2, g_3) in the absence of molecular symmetry, and at higher temperatures by two g values for cylindrical symmetry ($g_{\parallel}$ and $g_{\perp}$) due to rotation of the groups which results in averaging of the g values. This concept was used by many authors (Chien & Boss, 1967*a*; Moriuchi et al., 1969; 1970) in discussing the averaging of g anisotropy with motion and molecular symmetry. This problem has been further treated by Kashiwabara (1972)

The kinetics of macroradical oxidation can be determined from the ESR spectra of the peroxide radicals. In order to separate the overlapping spectra of the initial and the peroxide radicals, the experiments should be performed at relatively high level of microwave energy output. Under these conditions the saturation of the primary radical spectra occurs and the spectra of the peroxide radicals that remain far from saturation are observed on zero background (Bresler & Kazbekov, 1964). The macroradical oxidation (unless oxygen diffusion into the polymer influences the reaction) is given by the equation:

$$V_i = \ln \frac{N_\infty}{N_\infty - N} = kt$$

where N_∞ is the final number and N the number of peroxide radicals at time t and k is the rate constant. The kinetics of macroradical oxidation for three different polymers is given in Fig. 4.

A peroxy radical of a polymer can react with other polymer molecules and abstract a hydrogen atom:

$$PO_2^{\cdot} + PH \rightarrow POOH + P^{\cdot} \tag{26}$$

Peroxy radicals as a class are a relatively electrophilic species abstracting tertiary hydrogen in preference to secondary and primary hydrogens. Polymeric hydroperoxides are believed to be major products in the photooxidation of polymers (Adams, 1970; Heacock et al., 1968; Karjakin et al., 1955). On exposure to light they decompose to the polymeroxy radicals (PO·) and hydroxyl radicals:

$$POOH + h\nu \rightarrow PO^{\cdot} + HO^{\cdot} \tag{27}$$

The energy of the radiation of wavelength 360 nm is sufficiently high (78 kcal mol^{-1}) to cleave RO–OH bonds (42 kcal mol^{-1}) and probably also the R–OOH bond (70 kcal mol^{-1}) but not the ROO–H (90 kcal mol^{-1}) (Benson, 1965). These large differences in bond dissociation energy in the ROOH groups would mean that reaction (27) can be expected to be dominant. The photolysis of primary and secondary produces primary and secondary radicals which may further decompose by β scission:

$$
\begin{array}{c}
\quad R \qquad\quad R \\
\quad | \qquad\qquad | \\
-CH_2-C-CH_2-CH- \\
\quad | \\
\quad O
\end{array}
\rightarrow
\begin{array}{c}
\quad R \qquad\qquad R \\
\quad | \qquad\qquad | \\
-CH_2-C + {}^{\cdot}CH_2-CH- \\
\quad \| \\
\quad O
\end{array}
\tag{28}
$$

The β scission reaction is probably the most important cause of backbone scission during the photooxidation degradation of polypropylene, since about 15 % of the photolysed hydroperoxide groups are shown to follow this reaction (Carlsson & Wiles, 1969b).

The detection of RO· radicals by ESR spectroscopy has been reported (Piette et al., 1960) but disclaimed in later papers (Symons, 1963; Ingold & Morton, 1964).

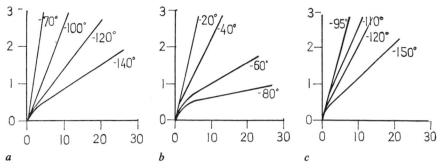

Fig. 4. *Abscissa:* t/min; *ordinate:* V_i.
Kinetics of macroradical oxidaton. *a*, polyvinylacetate; *b*, polymethyl methacrylate; *c*, polystyrene (Bresler & Kazbekov, 1964).

The Termination Reactions in Photooxidation

The termination of the reaction chain in photooxidation is due to reactions of free radicals with each other, in which inactive products are formed:

$$PO_2^\cdot + PO_2^\cdot \quad\text{―}$$ (29)

$$PO_2^\cdot + P^\cdot \quad\rightarrow \text{ inactive products}$$ (30)

$$P^\cdot + P^\cdot \quad\text{―}$$ (31)

The termination reaction is dependent on the oxygen pressure. When oxygen pressure is high the termination reaction almost exclusively follows equation (29). At low oxygen pressure not every termination will take place by the mutual reaction of peroxy radicals (29). Polymer radicals couple with themselves (31) and with peroxy radicals (30) forming crosslinks. Crosslinking and chain scission are secondary effects in the photooxidation reaction of polymers. Scission may turn a solid polymer into a liquid and crosslinking may give a brittle polymer network. Both these effects may occur in the same polymer, one at the surface, the other below the surface. The physical properties of the polymer may be strongly affected by these processes.

The Role of Structure of Macromolecules on Photooxidation of Polymers

There is an essential difference between the degradation and oxidation processes of solid polymers by UV radiation and the γ radiation. In the case of exposure to UV radiation free radicals are mainly produced on the surface, since even a thin layer of the polymer completely absorbs the UV light. γ Radiation, on the contrary, penetrates the whole bulk of the polymer and produces radicals everywhere (Voevodski, 1964).

In the case of polymer consisting of a crystalline and an amorphous phase, free radicals are on exposure to UV radiation chiefly formed in the amorphous phase, whereas the γ radiation produces them in the crystalline phase as well.

Many authors (Hansen, 1970; Hawkins et al., 1959; Luongo, 1963; Winslow et al., 1963; 1966) have reported that the rate of oxygen uptake during the thermal oxidation of solid polymers is inversely proportional to the percentage of crystallinity. Oxygen would only attack the amorphous part of the semicrystalline structure of polymers. Since the crystallinity is related to the tacticity of the polymers, stereoregular polymers are more resistant to oxidation than the atactic polymers (Hansen, 1970; Kato et al., 1969). Branching of the polymer molecules is another important factor in the oxidation process. Branched polyethylene (ethylene-butene copolymer) is more readily oxidized than the linear polymer (Hansen, 1970; Kondratiev et al., 1971; Winslow & Matreyek, 1964).

On account of the extremely large amount of theoretical and experimental material available on photooxidation of polymers, only some of the main problems have been discussed briefly in this paper.

These investigations are part of a research program supported by the Swedish Board for Technical Development and the Swedish Board of Building Research.

References

Adams, J. H., J. Polymer Sci., *8A1*, 1279 (1970).

Algar, B. E. & Stevens, B., J. Phys. Chem., *74*, 2728 (1970 *a*).

Algar, B. E. & Stevens, B., J. Phys. Chem., *74*, 3029 (1970 *b*).

Amerik, Y. & Guillet, J. E., Macromolecules, *4*, 375 (1971).

Benson, S. W., J. Chem. Educ., *42*, 502 (1965).

Bolland, J. L., Quart. Rev., *3*, 1 (1949).

Bourdon, J. & Schuringer, B., in Physics and Chemistry of the Organic Solid State (ed. D. Fox, M. M. Labes & A. Weissberger) vol. 3, chapt. 2. Interscience, New York 1967.

Bresler, S. E. & Kazbekov, E. N., Fortschr. Hochpolym. Forsch., *3*, 688 (1964).

Carlsson, D. J., Kato, Y. & Wiles, D. M., Macromolecules, *1*, 459 (1968).

Carlsson, D. J. & Robb, J. C., Trans. Faraday Soc., *62*, 3403 (1966).

Carlsson, D. J. & Wiles, D. M., Macromolecules, *2*, 587 (1969 *a*).

Carlsson, D. J. & Wiles, D. M., Macromolecules, *2*, 597 (1969 *b*).

Chien, J. C. W. & Boss, C. R., J. Am. Chem. Soc., *89*, 571 (1967 *a*).

Chien, J. C. W. & Boss, C. R., J. Polymer Sci., *5A1*, 3091 (1967 *b*).

Corey, E. J. & Taylor, W. C., J. Am. Chem. Soc., *86*, 3881 (1964).

Cowell, G. W. & Pitts, J. N. Jr., J. Am. Chem. Soc., *90*, 1106 (1968).

Day, M. & Wiles, D. M., J. Appl. Polymer Sci., *16*, 203 (1972).

Ershov, Yu., Kinet. Kataliz., *10*, 577 (1969).

Ershov, Yu., Kuzmina, S. I. & Neiman, M. B., Uspkh. Khim., *38*, 289 (1969).

Fallick, A. M., Mahan, B. H. & Meyers, R. J., J. Chem. Phys., *42*, 1837 (1965).

Fox, R. B. & Priece, T. R., J. Appl. Polymer Sci., *11*, 2373 (1967).

Glaeson, W. S., Rosenthal, I. & Pitts, J. N. Jr., J. Am. Chem. Soc., *92*, 7042 (1970).

Gollnick, K., Adv. Photochem., *6*, 2 (1968 *a*).

Gollnick, K., Adv. Chem. Ser., No. 77, 67 (1968*b*).

Guillet, J. E., Dhanraj, J., Golemba, F. J. & Hartley, G. H., Adv. Chem. Ser., No. 85, 272 (1968).

Guillet, J. E. & Norrish, R. G. W., Proc. Roy. Soc., *A233*, 153 (1955).

Hansen, R. H., in Thermal Stability of Polymers (ed. R. T. Conley) p. 153. Dekker, New York, 1970.

Hama, Y., Hosono, K., Furui, Y. & Shinohara, K., J. Polymer Sci., *9A1*, 1411 (1971).

Hartley, G. H. & Guillet, J. E., Macromolecules, *1*, 165 (1968*a*).

Hartley, G. H. & Guillet, J. E., Macromolecules, *1*, 413 (1968*b*).

Hawkins, W. L., Matreyek, W. & Winslow, F. H., J. Polymer Sci., *41*, 1 (1959).

Heacock, J. F., Mallory, F. B. & Gay, F. P., J. Polymer Sci., *6A1*, 2921 (1968).

Heskins, M. & Guillet, J. E., Macromolecules, *3*, 224 (1970).

Ingold, K. U. & Morton, J. R., J. Am. Chem. Soc., *86*, 3400 (1964).

Iwasaki, M. & Sakai, Y., J. Polymer Sci., *6A2*, 265 (1968).

Joffe, Z. & Rånby, B, This volume, p. 171 (1972).

Kaplan, M. L. & Kelleher, P. G., J. Polymer Sci., *8A1*, 3163 (1970).

Kaplan, M. L. & Kelleher, P. G., J. Polymer Sci., *9B*, 565 (1971).

Karyakin, A. V., Nikitin, V. A. & Sidorov, A. N., Zhur. Fiz. Khim., *29*, 1624 (1955).

Kashiwabara, This volume, p. 275 (1972).

Kato, Y., Carlsson, D. J. & Wiles, D. M., J. Appl. Polymer Sci., *13*, 1447 (1969).

Kearns, D. R., Prepr. Div. Petrol. Chem. Am. Chem. Soc., *16*, A9 (1971).

Kearns, D. R., Hollins, R. A., Khan, A. U., Chambers, R. W. & Radlick, P., J. Am. Chem. Soc., *89*, 5455 (1967*a*).

Kearns, D. R., Hollins, R. A., Khan, A. U. & Radlick, P., J. Am. Chem. Soc., *89*, 5456 (1967*b*).

Kearns, D. R., Khan, A. U., Duncan, C. K. & Maki, A. H., J. Am. Chem. Soc., *91*, 1039 (1969).

Kondratiev, V. I., Sisin, A. A. & Tamaev, E. E., Plast. Massy, *75*, 48 (1971).

Kummler, R. H. & Bortner, M. H., Environ. Sci. Technol., *3*, 944 (1969).

Kwei, K. P. S., J. Polymer Sci., *7A1*, 1075 (1969).

Lukac, I., Hrdlovic, P., Manasek, Z. & Bellus, D., J. Polymer Sci., *9A1*, 69 (1971).

Luongo, J. P., J. Polymer Sci., *1B*, 141 (1963).

Monahan, A. R., J. Polymer Sci., *4A1*, 2381 (1966).

Mönig, H., in Probleme und Ergebnisse aus Biophysik und Strahlenbiologie, vol. 3, p. 55. Akademie Verlag, Berlin, 1962.

Mönig, H. & Kriegel, H., Z. Naturforsch., *15b*, 333 (1960).

Moriuchi, S., Kashibawara, H., Shoma, J. & Yamagushi, S., J. Chem. Phys., *51*, 2981 (1969).

Moriuchi, S., Nakamura, M., Shimada, S., Kashibawara, H. & Shoma, J., Polymer, *11*, 630 (1970).

Noda, S., Fueki, K. & Kuri, Z., J. Chem. Phys., *49*, 3287 (1968).

Norling, P. M. & Tobolsky, A. V., in Thermal Stability of Polymers (ed. R. T. Conley) p. 113. Dekker, New York 1970.

Partridge, R. H., J. Chem. Phys., *45*, 1679 (1966).

Partridge, R. H., J. Chem. Phys., *49*, 3656 (1968).

Piette, L. H. & Landgraf, W. C., J. Chem. Phys., *32*, 1107 (1960).

Rabek, J. F., Chemia Stosow., *11*, 53 (1967*a*).

Rabek, J. F., Chemia Stosow., *11*, 73 (1967*b*).

Rabek, J. F., Chemia Stosow., *11*, 89 (1967*c*).

Rabek, J. F., Chemia Stosow., *11*, 183 (1967*d*).

Rabek, J. F., Photochem. Photobiol., *7*, 5 (1968*a*).

Rabek, J. F., Research Report, Uppsala University, Sweden (1968*b*).

Rabek, J. F., Wiadom. Chem., *25*, 293, 365, 435 (1971*a*).

Rabek, J. F., Polimery–Tw. Wielkocz., *16*, 257 (1971 *b*).

Rabek, J. F., Pure Appl. Chem., *8*, 29 (1971 *c*).

Rabek, J. F., Comprehensive Chemical Kinetics (ed. C. H. Bamford). In press.

Rabek, J. F. & Rånby, B., Lecture presented at IUPAC Symposium on Photochemical Processes in Polymer Chemistry, Louven, Belgium, June 12–15, 1972.

Rabek, J. F. & Skorupa, L., Unpublished results (1971).

Rånby, B. & Yoshida, H., J. Polymer Sci., Part C, No. 23, 263 (1966).

Siegel, S. & Hedgpeth, H., J. Chem. Phys., *46*, 3904 (1967).

Siewert, H., Z. Naturforsch., *19 b*, 806 (1964).

Star, R., Sprung, J. & Pitts, J. N. Jr., Environ. Sci. Technol., *3*, 946 (1969).

Stauff, J. & Fuhr, H., Ber. Bunsenges. Phys. Chem., *73*, 245 (1969).

Stevens, B. & Algar, B. E., J. Phys. Chem., *72*, 2582, 2468, 3794 (1968).

Stevens, B. & Algar, B. E., J. Phys. Chem., *73*, 1711 (1969).

Symons, M. C. R., Adv. Phys. Organ. Chem., *3*, 307 (1963).

Trozzolo, A. M., Adv. Chem. Ser., No. 77, 167 (1968).

Trozzolo, A. M. & Winslow, F. H., Macromolecules, *1*, 98 (1968).

Tryon, M. & Wall, L. A., in Autooxidation and Antioxidants (ed. L. Lundberg) p. 919. Interscience, New York 1962.

Tsuji, K. & Seiki, T., Polymer J., *1*, 133 (1970 *a*).

Tsuji, K. & Seiki, T., J. Polymer Sci., *8B*, 817 (1970 *b*).

Tsuji, K. & Seiki, T., J. Polymer Sci., *9A1*, 306 (1971 *a*).

Tsuji, K. & Seiki, T., Polymer J., *2*, 606 (1971 *b*).

Tsuji, K. & Seiki, T., Reports Progr. Polymer Phys., Japan, *14*, 577 (1971 *c*).

Tsuji, K. & Seiki, T., Reports Progr. Polymer Phys., Japan, *14*, 581 (1971 *d*).

Tsuji, K. & Seiki, T., Reports Progr. Polymer Phys., Japan, *14*, 585 (1971 *e*).

Tsuji, K. & Seiki, T., Reports Progr. Polymer Phys., Japan, *14*, 589 (1971 *f*).

Voevodski, V. V., Proc. Tihany Sympos., Akademii Kiado, Budapest, 1964, p. 112.

Wasserman, E., Kuck, V. J., Delevan, W. M. & Yager, W. M., J. Am. Chem. Soc., *91*, 1040 (1968).

Wilson, T., J. Am. Chem. Soc., *88*, 2898 (1966).

Winslow, F. H., Aloisio, C. J., Hawkins, W. L., Matreyek, W. & Matsuoka, S., Am. Chem. Soc., Polymer Preprints, *4*, 706 (1963).

Winslow, F. H., Hellman, M. Y., Matreyek, W. & Stills, S. M., Polymer Engl. Sci., *6*, 273 (1966).

Winslow, F. H. & Matreyek, W., Am. Chem. Soc., Polymer Preprints, *5*, 552 (1964).

Yoshida, H. & Rånby, B., J. Polymer Sci., *2B*, 1155 (1964).

Reaction of Free Radicals with Electrons during Degradation of Polymers

By Péter Hedvig

Research Institute for Plastics, Budapest, Hungary

The degradation of polymers is generally considered as being a free radical process. Difficulties have arisen, however, in understanding how the thermal process is initiated and why many polymers are sensitive to low energy illumination without exhibiting absorption band in the corresponding energy range. The situation is complicated with regard to the effect of the physical structure of the polymer to the chemical reaction. On the other hand some experimental evidence has been collected about the presence of a considerable concentration of charge carriers (electrons and holes) trapped in polymeric solids. These carriers can be re-mobilized, detrapped, by thermal and/or by radiation energy. The charge carriers should be chemically active either directly with electro or nucleophilic processes or indirectly by forming excitons by recombination. In each case, the only thing needed for creating chemically active species is the energy of detrapping carriers, which is in the order of 0.1 eV, according to thermostimulated current and photoconductivity measurements. The author believes that thermal degradation processes and those induced by visible light in solid polymers are initiated by detrapped charge carriers.

Detrapped charge carriers should be chemically active also in such processes in which the energy is sufficient to create free radicals directly by bond scissions, as by UV, X-ray or γ-ray irradiation. As during irradiation, charge carriers are besides free radicals, also activated. The free-radical yields measured after irradiation are not necessarily characteristic to the probability of the corresponding bond scissions, but are rather results of equilibrium between processes involving formation, trapping and reactions of radicals and carriers.

With respect to this it is essential to see if there is experimental evidence supporting the possibilities of reactions between free radicals and charge carriers in polymeric solids. This problem has been studied by combining ESR technique with radiation-induced conductivity, photoconductivity and thermally stimulated current measurements. As detrapping of charge carriers is considerably dependent on the physical structure of the polymer, some structural studies have also been performed, especially mechanical and dielectric relaxation, differential scanning calorimetry and wide angle X-ray diffraction measurements. The main results are summarized subsequently.

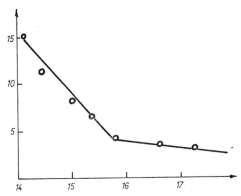

Fig. 1. *Abscissa:* log radical conc. (spin/g); *ordinate:* induced conductivity. $10^{12}/\Omega^{-1}$ cm^{-1}.
Dependence of the radiation-induced conductivity on the trapped free radical concentration in LDPE. Dose rate 60 Mrad h^{-1}; pulse length for conductivity measurement 10 ms. Temperature of irradiation and measurement 20°C.

ESR and Radiation Induced Conductivity Studies

Radiation-induced conductivity and ESR measurements were performed on polyethylene, polytetrafluoroethylene, polychlorotrifluoroethylene, polystyrene, polyethylene terephthalate and polycaprolactam by using sequences of pulsed and continuous X irradiation. Induced conductivities were recorded in the pulsed (1 ms–1 min) regime, and the ESR signals were measured after certain irradiation periods (10–100 min). In this way induced conductivities of samples containing known concentrations of trapped free radicals could be measured. Under such conditions the induced conductivity was found to decrease by increasing radical concentration. This is illustrated in Fig. 1 for LDPE irradiated and measured at 20°C. Upon storage at the same temperature or upon annealing at somewhat elevated temperatures the induced conductivity was found to increase along with the decay of the radical concentration.

By using radical scavengers such as hydroquinone the preirradiation effect could be eliminated; in such cases no trapped radicals could be detected by ESR.

The Stepwise Decay of Radicals

The decay of radicals trapped after irradiation was measured at stepwise increasing temperatures for polyethylene, polytetrafluoroethylene, polychlorotrifluoroethylene and polyethylene terephthalate. The decay curves exhibited fast and slow portions: at each temperature a considerable concentration of radicals became strongly trapped in the solid; the decay was practically stopped. A typical decay curve for peroxy radicals in polychlorotrifluoroethylene

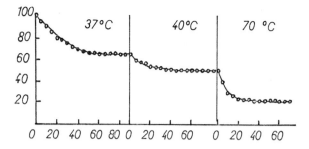

Fig. 2. *Abscissa:* time (min); *ordinate:* [ROO] %.
Stepwise decay of peroxy radicals in polychlorotrifluoroethylene. Radiation dose: 20 Mrad at 20°C.

is shown in Fig. 2. For comparison several organic and anorganic non-polymeric solids were also studied; the stepwise character of the radical decay was observed in each case (Voevodsky, 1964; Hedvig, 1968; Tolkachev, 1972).

By comparing the results with molecular mobility measurements performed by dielectric and mechanical relaxation methods it was concluded that the stepwise character of the radical decay could not be interpreted by a bimolecular radical–radical recombination process. By considering the possibility of reactions of trapped radicals with thermally mobilized electrons the decay curves could be satisfactorily described by eq. (1) (Hedvig, 1969 *a*),

$$\frac{dR}{R} = kR[e_0 - (R_0 - R)]\,dt \tag{1}$$

where R_0, R are the total trapped, respectively instantaneous radical concentrations, e_0 is the total concentration of trapped electrons, k is the rate constant of the process. If it is assumed that the radical decay is entirely due to reactions with detrapped charge carriers the temperature dependence of the radical decay steps should represent the energy distribution of the trapped carriers. In Fig. 3 these steps are illustrated as a function of the temperature in an Arrhenius system. It is seen that a fair straight line is obtained; the apparent activation energy is 0.12 eV. This should correspond to the average energy of the traps.

Detection of Chain-End Radicals Formed by Radiation Degradation

A lot of experimental evidence has been collected so far by means of ESR about structural changes of trapped polymeric radicals upon irradiation, sometimes even upon illumination with visible light. Some thermally induced transformations are also known (see e.g. Ayscough, 1967). For the study of degradation processes structural transformations of radicals are evidently of

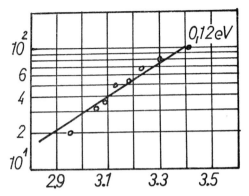

Fig. 3. *Abscissa:* $1\,000/T^{-1}/K^{-1}$; *ordinate:* $\Delta R/R\,\%$.
   Arrhenius plot of the radical decay steps for obtaining overall activation energy for carrier detrapping: 0.12 eV. Sample: polychlorotrifluoroethylene, preirradiation dose: 20 Mrad at 20°C.

basic importance. During radical degradation processes involving decrease in molecular weight chain-end radicals as

$$R\!=\!\!-\!\overset{|}{\underset{|}{C}}\!-\!\overset{|}{\underset{|}{C}}\!-\!\overset{|}{\underset{|}{C}}\!\cdot$$

should appear. Nevertheless after irradiation or UV illumination of polymers only chain-side radicals

$$R\!=\!\!-\!\overset{|}{\underset{|}{C}}\!-\!\overset{|}{\underset{}{\dot{C}}}\!-\!\overset{|}{\underset{|}{C}}\!-$$

can be detected. Chain-end radicals could only be detected so far by mechanical destruction (Butyagin, 1961), by chemomechanical action (Dobo & Hedvig, 1967) or by combined high energy-low energy irradiation and oxidation (Hedvig, 1969*b*) by the following process

$$P \xrightarrow{h\nu_1} \dot{R} \xrightarrow{O_2} RO\dot{O} \xrightarrow{h\nu_2} R^\cdot \qquad (2)$$

where P means a polymer molecule, $h\nu_1$ is X-irradiation energy, $h\nu_2$ UV irradiation energy.

   It has been shown by ESR studies that chain-end radicals stabilized either way would transform quickly upon repeated irradiation as follows

$$R^\cdot \xrightarrow{h\nu_1} \dot{R} \qquad (3)$$

   This reaction has been found to be very fast even at low temperatures. Chain-end radicals could be stabilized so far only in polytetrafluoroethylene by process (2). In other polymers, such as polyethylene and polypropylene, the chain-end radicals could not even be stabilized at $-196°C$. Only Rånby (1967) has found some evidence of a small portion of chain-end radicals at $-196°C$ after UV illumination in polyethylene.

From ESR studies an important conclusion could be drawn about radiation-induced oxidative degradation. Since the actual chain-scissions occur at the peroxy radicals and reaction (3) is very fast in the radiation field, the total process is

$$\text{non radical products}$$

$$P + h\nu_1 \rightarrow \left\{ k_1 \left| \begin{array}{ccc} \uparrow & \uparrow & \uparrow \\ k_4 & k_5 & k_6 \\ \downarrow & \downarrow & \downarrow \\ R \xrightarrow[k_2]{O_2} RO\dot{O} \xrightarrow[k_3]{h\nu_2} R\cdot \\ \underset{}{\nwarrow} \underset{k_1}{} \nearrow \\ R\cdot \end{array} \right. \right. \tag{4}$$

where $h\nu_1, h\nu_2$ mean high energy respectively low energy (UV) quanta, $k_1 \ldots k_6$ are the rate constants of the corresponding reactions; $k_4, k_5, \ll k_2, k_3 \ll k_1$

From eq. (4) it is seen that the overall efficiency of the radiation depends on the rate of oxidation (k_2) as well as on the stability of the chain-end radicals (k_6) at the given temperature. The situation is somewhat paradox, as those polymers would degrade faster for which the chain-end radicals are most stable (e.g. PTFE).

It is believed that process (4) holds for any non cross-linked polymer, although it has only been verified by ESR for PTFE only. Some indirect evidence by analyzing radical yields and by studying mechanical properties has been found for polyethylene and for polypropylene as well (Hedvig, 1969c).

Although the radiation-induced oxidative degradation process has become rather clear by the ESR studies it remains questionable whether the very fast reactions k_1 and k_3 are due to direct excitations or to indirect effects. The reaction k_3 in PTFE has been studied by using filtered UV illumination and found not to be selective. This would mean that the peroxy group is not directly excited by the radiation energy to result in chain scissions. As has been shown previously in this paper there is a possibility for the peroxy radicals to react with radiation-activated electrons to produce excited ions which, in turn, would dissociate to result in chain scissions. This process is theoretically possible, although no direct experimental proof is available, since the intermediate excited ions are difficult to detect.

Thermostimulated Current Measurements and Problems of Structural Transitions

The ESR method is mainly used for studying radiation-induced degradation processes because by thermal degradation the concentration of radicals is usually too small to be detected. Some information about the thermal process, however, can be gained by studying the thermal decay of radicals trapped after previous irradiation. As has been discussed earlier in this paper, the stepwise decay of trapped radicals cannot be interpreted in terms of molecular mobilities, because in those temperature ranges where the decay is observed, translational mobilities are practically frozen-in. One should either assume that radicals would migrate along the polymer chains (Dole & Keeling, 1953), or that radicals would not actually recombine but would react with other species as electrons and holes detrapped by thermal energy.

In order to show that fairly high concentrations of charge carriers are present in solid polymers thermally stimulated current (for review, see van Turnhout, 1971) measurements have been performed and the effect of preirradiation on these current spectra has been studied. By these experiments the thermal history of the samples was carefully fixed, and the results were compared with those of mechanical and dielectric relaxation experiments. Three types of such experiments have been performed:

(a) measurement of the current as a function of the temperature without external voltage for a thermomechanically pretreated sample
(b) for a sample polarized under an electric field above a structural transition temperature
(c) for a sample irradiated at low temperature.

Measurements of type (a) resulted in faint current peaks representing the main structural transitions of the polymer. By the method (b) these peaks appeared very intense. By irradiation on (a)-type spectra intense thermostimulated current peaks are superimposed: These peaks do not correspond to the structural transitions. The results are illustrated in Fig. 4 for LDPE. Curve (1) corresponds to case (c) showing the background of the untreated sample and the effect of irradiation at $-196°C$. Curve (2) corresponds to a spectrum measured after polarization at $100°C$ by an electric field of $10 \, kV \, cm^{-1}$. Curve (3) is a mechanical relaxation spectrum measured at a frequency of 8 Hz by using identical temperature program as by the depolarization measurements.

From the thermostimulated current measurements it is concluded that even in untreated polymers a considerable concentration of trapped charge carriers are present. Assuming that the drift mobility in polymers is in the order of $10^{-3} \, V \, cm^{-1}$ it is concluded that this concentration is about 10^{16} carriers g^{-1}.

Another important conclusion is that the carriers trapped after irradiation are energetically unevenly distributed, i.e. it is possible to activate considerable concentrations of carriers much below the structural transition temperatures. There are principally two basic ways for detrapping the carriers: one is direct excitation without disturbing the trapping sites, the other is destruction of the

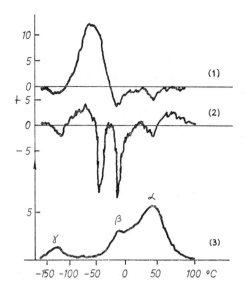

Fig. 4. *Abscissa:* temperature (°C); *ordinate:* (*top*) current × 10^{-11}/A; (*bottom*) mechanical loss, log G″.

Thermostimulated current spectra for LDPE. (1) after irradiation at −196°C with X rays (2 Mrad); (2) after polarization at −100°C under an electric field of 10 kV cm⁻¹ for 1 h. Curve (3) represents a mechanical relaxation spectrum at 8 Hz. Rates of heating and cooling: 2°C min⁻¹ for all curves.

trapping sites (defects) themselves at structural transitions, especially at glass-rubber transition.

By analysing the thermostimulated current curves it is possible to obtain approximate activation energy values for charge detrapping. This figure is in the order of 0.1 eV for most polymers. For polyethylene, for example, below the α transition the detrapping activation energy is found to be 0.4 eV (between −150 and −20°C).

References

Ayscough, P. B., Electron Spin Resonance in Chemistry. Methuen, London 1967.
Butyagin, P. Yu., Dokl. Akad. Nauk. SSSR, *140*, 145 (1961).
Dobó, J. & Hedvig, P., J. Polymer Sci. C, *16*, 2577 (1967).
Dole, M. & Keeling, C. D., J. Amer. Chem. Soc., *75*, 6082 (1953).
Hedvig, P., J. Polymer Sci. A.2., 4097 (1964).
Hedvig, P., Eur. Polymer J., Suppl., p. 285 (1969*a*).
Hedvig, P., J. Polymer Sci., *A7*, 1145 (1969*b*).
Hedvig, P., Kinetics and Mechanisms of Polyreactions. Proc. IUPAC Conference, Hung. Acad. Sci., Budapest, *5*, 277 (1969*c*).
Rånby, B. & Yoshida, H., J. Polymer Sci. C, *12*, 263 (1966).
Tolkachev, V., Radiation Chemistry. Proc. 3rd Symposium, Tihany, Hungary (ed. J. Dobó & P. Hedvig). Publ. House of Acad. Sci, Hungary 1972.
Turnhout, van, J., Polymer J., *2*, 173 (1971).
Voevodsky, V. V., Radiation Chemistry. Proc. 1st Tihany Symposium (ed. J. Dobó) Publ. House of the Acad. Sci., Hungary 1964.

Discussion

Yoshida

When the trapped charge carriers are released thermally, they are expected to move to random directions. As a whole, the electric current is not expected to be observed without applied electric field, even though the charge carriers are non-uniformly trapped microscopically before their release. What is your interpretation of the thermostimulated currents you observed?

Hedvig

There is a persistent internal polarization in the bulk of the polymer after polarization by electric field and after irradiation as well. This means that the polymer behaves like an electret. Moreover, any change in the internal distribution of charge density should result in a current according to the laws of electrodynamics. So the charges do not need to diffuse through the polymer in order to produce current (see e.g. Freeman, Kallmann & Silver, Rev. Mod. Phys., *33*, 553 (1961)).

In the particular case of polymers I think that there are two different mechanisms: detrapping of carriers which produce thermostimulated current and destruction of Maxwell–Wagner type dipoles which produce depolarization peaks.

Lindberg

Have you tried to investigate the phenomena discussed in strongly semiconductive materials?

Hedvig

We did not try to study such systems. As far as I know nobody has seen radicals in irradiated organic semiconductors. The reason for this might be the reaction of radicals with electrons.

Sohma

Why did you assign the two peaks appearing in the β region to the melting of the frozen dipole? It seems to me that only small scale of the molecular motion, such as hindered rotation, may be enough to reduce the polarization. If so, the molecular motion of large scale, like the glass transition is not required for the depolarization of the frozen polarization.

Hedvig

I think that this is not the usual thermally induced depolarization, but that it is due to the destruction of the structural defects themselves. This involves that the β transition in low density poyethylene is a kind of structural transition in which the entropy and free volume changes. This is indicated by DSC measurements and by the sensitivity of the amplitude of the depolarization peak to previous thermal treatment. On the basis of this I think that in the β region rather drastic structural rearrangement takes place.

MECHANICAL FRACTURE OF POLYMERS

ESR Studies on the Polymer Radicals Produced by Mechanical Fracture

By J. Sohma, T. Kawashima,[1] S. Shimada,[2]
H. Kashiwabara[2] and M. Sakaguchi

College of Engineering, Hokkaido University, Sapporo, Japan

Introduction

It was established by the several works (Butyagin, Kolbanev & Radtsig, 1964; Campbell & Peterlin, 1968; Peterlin, 1971; Sohma et al. 1969; DeVries, 1971; Kausch, 1970) that ESR is a useful experimental technique to study the molecular mechanisms of either mechanical fracture or large deformation by detecting free radicals produced by fracture or deformation. On the other hand, there are a lot of ESR studies (Campbell, 1970) on polymer radicals produced by ionizing radiation and UV irradiation. It seems to be interesting to compare the radicals produced by the two different methods and to make trials for finding out any clues for the mechanisms of mechanical fracture, as well as the irradiation effect.

Method of Mechanical Fracture

In our experiments the mechanical sawing was adopted for the method of mechanical fracture. The polymer samples were sawed in liquid nitrogen in a vessel and the saw dusts of the polymer produced by sawing were transferred with liquid nitrogen to a sample tube. This tube, containing saw dusts and liquid N_2, was connected to the vacuum system and evacuated, keeping the sample tube in the dewar containing liquid N_2. By this procedure the saw dusts of a polymer were produced at 77 K and preserved at the same temperature until warming experiments. It should be mentioned that the samples were not in contact with oxygen except the impurity in liquid N_2.

Radicals Produced by the Mechanical Fracture

Identification of the Radicals

(a) *Polyethylene*: The spectrum (Fig. 1a) of the radicals primarily produced by the mechanical fracture is different from that of the radical produced by

Present address: [1] Nissan Chemical Co.; [2] Nagoya Institute of Technology, Hokiso, Showa-Ku, Nagoya 466, Japan.

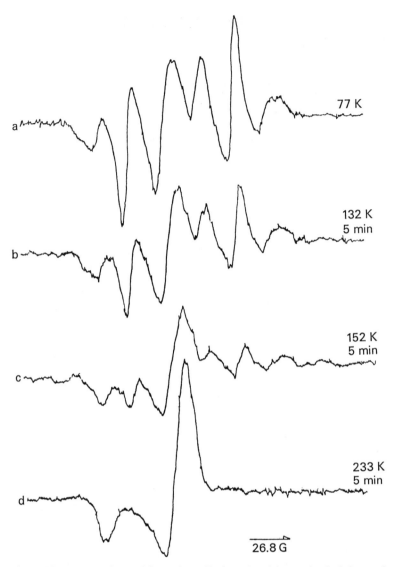

Fig. 1. ESR spectra observed from the radical produced by mechanical destruction of poly-ethylene. *a*, 77 K; *b*, 132 K, 5 min; *c*, 152 K, 5 min; *d*, 233 K, 5 min.

γ irradiation. The radical responsible for the spectrum was identified as a
scission type radical, $\cdot\text{C-C-}$, as shown in Fig. 2. This assignment was confirmed
by the fact that the spectrum of the radical $\cdot\text{C-C-CH}$ obtained from BrC_3H_7
was identical to that assigned to the scission radical, as is also shown in Fig.
2. The predominant component of the spectrum observed at 77 K after the heat

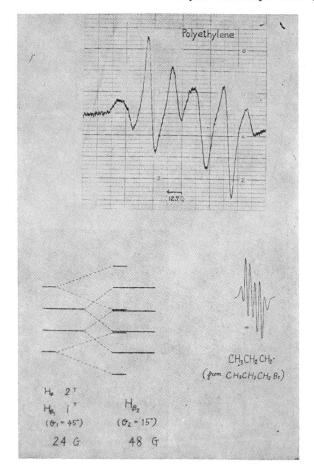

Fig. 2. Assignment of the spectrum obtained in the mechanical fracture.

treatment at 152 K for 5 min (Fig. 1c) was the octet with the separation of 23G, which was identified as the spectrum from the radical,

$$\begin{array}{ccc} & H & H & H \\ H-C-C-C & \\ & H & \cdot & H \end{array}$$

(Shimada, Kashiwabara & Sohma, 1970). This conversion of the radical species demonstrates the hydrogen abstraction from the adjacent site in the same molecule. The spectra obtained after the heat treatment (Fig. 1d) was the characteristic spectrum of the peroxyradical, presumably

$$\begin{array}{cc} & H & H \\ H_3C - C - C - \\ & OO^{\cdot} & H \end{array}$$

in this case.

(b) *Polypropylene:* The spectrum observed immediately after the fracture was the mixture of the spectrum of the peroxy radical and a multiplet (Fig. 3a). From the survived half of the multiplet one could determine the multiplet as the octet having the separation of 23 G. These analyses convinced us that the radicals primarily produced by the mechanical fracture was the scission type

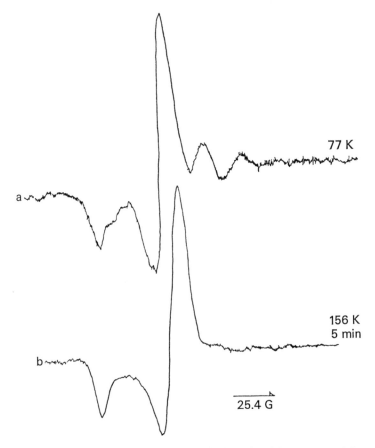

Fig. 3. ESR spectra observed from the radical produced by mechanical destruction of polypropylene. *a*, 77 K; *b*, 156 K, 5 min.

CH₃ H
·C – C–. By heat treatment the radicals converted into peroxyradicals, as shown
H   H
in Fig. 3*b*.

(c) *Polytetrafluoroethylene:* The spectrum observed immediately after the mechanical fracture was the peroxy radical type in the case of polytetrafluoroethylene. Although it was hard to say whether or not the scission radical was produced from the line shape, the analysis of the temperature variation of the spectrum leads to the conclusion that the radical produced by the mechanical

F F
fracture was the scission type radical, ·O–O–C–C– (Sohma et al., 1970).
F F

Characteristics of the Polymer Radical Produced by the Mechanical Fracture

(a) The radical produced by mechanical fracture of the polymers were mostly the chain-scission radical. On the other hand, the species of the polymer radicals produced by ionizing radiation are mostly the chain radicals.

(b) No ESR spectrum was observed from HNO_3-treated polyethylene and the solid of the low molecular compounds, such as benzene, ethanol, and paraffin, after the mechanical fracture by the same method at 77 K. These results indicate that creation of the radicals, that is the bond fracture, is the characteristic property of polymers, the length of which is probably more than 100 Å.

(c) The radicals produced by the mechanical fracture were easily converted into peroxy radicals, although the oxygen content in the systems was as small as the impurities in either liquid nitrogen or the original polymers. Such an easy conversion to the peroxy radical was not observed for the radicals produced by radiation. Thus, the high reactivity to oxygen is the characteristic of radicals produced mechanically. This difference of reactivity seems to suggest the difference in the trapping sites of the radicals depending on two different methods of production; the radicals produced by mechanical destruction are on the fresh surface, while radicals produced by irradiation are mostly deep in the samples. In Table 1 the species identified for the radicals produced by the mechanical fracture were listed for several polymers.

The Decay Curves

(a) *Polypropylene:* The relative intensities of the spectra observed at 77 K after heat treatment were plotted against the temperatures at which the sample had

Table 1.

Samples	Probable free radicals	Samples	Probable free radicals
Teflon	CF_2–$CF_2OO^{\cdot}$	Polyoxymethylene	$-CH_2O^{\cdot}$
Polyethylene	$-CH_2$–$CH_2{}^{\cdot}$, H_3C-C-C- and $ROO^{\cdot}$ (with H H above, $^{\cdot}$ H below)	Polymethyl-methacrylate	Mainly $-CH_2$–$C^{\cdot}$ (with CH_3 above, $COOCH_2$ below)
Polypropylene	$^{\cdot}C$–C- (with CH_3 H above, H H below) $ROO^{\cdot}$	HNO_3-treated polyethylene	No signal observed (bearing-ball method)
Polybutadiene (*cis*-rich)	$ROO^{\cdot}$ (spectrum is superposed with quartet spectrum)	Paraffin	No signal observed
Polybutadiene (54 % *trans*)	$ROO^{\cdot}$	Benzene, ethanol (several compounds with low molecular weights)	No signal observed

230   *J. Sohma et al.*

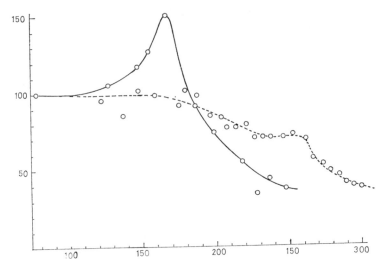

Fig. 4. *Abscissa: T/K; ordinate:* % rel. intensity.
Decay curves of polypropylene mechanical fracture (–○–) and γ irradiation (--○--).

been heat-treated for 5 min. This curve is called the decay curve. The decay curves were reproduced in Fig. 4 for comparison of the two cases—the mechanical fracture and γ irradiation. ESR spectra observed in the case of mechanical fracture were found to be enhanced by heat treatment at elevated temperatures instead of the common behavior (Nara, Kashiwabara & Sohma, 1967) of decrease of the spectral intensity. It was checked that this increase in ESR intensity was not originated either from ESR desaturation or line-narrowing. Thus, the increase of intensity observed in this temperature region should be considered due to increase of the radical concentration. The other characteristics of the decay behavior of the mechanically produced radicals is the rapid decay in the temperature region above 200 K.

(b) *Polytetrafluoroethylene:* In this case also the increase of the radical concentration was observed at elevated temperatures, while no increase in the decay curve was found for the radical produced by γ irradiation. The radicals produced mechanically decays rapidly at higher temperatures.

(c) *Polyethylene:* Although the polyethylene radicals formed by destruction decay in three steps as the radicals produced by irradiation (Shimada, Kashiwabara & Sohma, 1968), a similar increase of the radical concentration was found in the second stage of the decay curve. The mechanically produced radicals decay more rapidly in the third stage of the decay curve in this case as well.

Some Factors Affecting the Decay Behaviors of the Radicals Produced by Mechanical Destruction

Effect of Oxygen

The effect of oxygen was checked by controlling the pressure of oxygen in the sample tube. It was found that the increase of ESR intensities in the temperature range was enhanced with increased pressure or oxygen. This fact indicates that oxygen contributes to the formation of new radicals in this temperature range, at least in the case of polytetrafluoroethylene.

Effect of the Electric Change

The decay curves were obtained for the same samples in two different conditions of experiments: (1) the whole apparatus including experimentalist was electrically insulated with great care; (2) saw dusts of sample are not insulated but electrically grounded. In Fig. 5 the decay curves are shown for polypropylene in different experimental conditions. The decay curve obtained by the insulated experiment is quite different from that in non-insulation. The radicals survived in the case of electrical insulation in the higher temperature range, 150 K– 300 K. On the other hand, in the noninsulated experiment the radicals decayed more rapidly in this temperature range. It was found that oxygen has the effect to increase the radical concentration. In insulation experiments of the decay curve of polypropylene radicals, oxygen showed that the effect reduces the amount of radical increase, which was an opposite effect to the oxygen

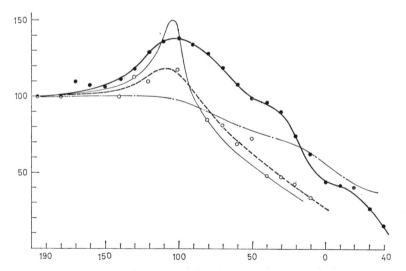

Fig. 5. *Abscissa: t/°C; ordinate:* % rel. intensity.
Decay curves of polypropylene. Mechanical fracture. —●—,⸢insulation; --○--, non insulation; decay curve taken without care for insulation, the same as the solid line in Fig. 4; —·—·—, γ-irradiation, the same curve as the dotted line in Fig. 4. Pressure 10^{-4} mm Hg.

effect for polytetrafluoroethylene mentioned on p. 228. Although these are preliminary results the electric charge seems to have some effect on the mechanism of decay, as well as on the formation of radicals at elevated temperatures.

Conclusions

Difference in the Species

The radical species produced by mechanical destruction are radicals formed by chain scission, while radicals produced by γ irradiation are not of scission type, but chain radicals.

Reactivity to Oxygen

Radicals produced by mechanical destruction are rather reactive with oxygen and easily formed peroxy radicals. On the other hand, radicals formed by γ irradiation are not so reactive to oxygen. This difference of reactivity suggests a difference in the sites where radicals were formed and trapped.

Increase of Radicals at Elevated Temperatures

In the case of mechanical destruction new radicals are produced at elevated temperatures, while radicals merely decay with increased temperatures in case of γ irradiation.

Effects of Oxygen

Oxygen has the effect to enhance radical formation at elevated temperatures in the case of polytetrafluoroethylene radicals produced by mechanical destruction.

Effect of Electric Charge

The electric charge formed in the process of mechanical destruction has some effects on the decay and formation of radicals of polypropylene.

References

Butyagin, P. Yu., Kolbanev, I. V. & Radtsig, V. A., Soviet Physics, Solid State, *5*, 1642 (1964).
Campbell, D. & Peterlin, A., Polymer Letters, *6*, 481 (1968).
Campbell, D., Macromol. Rev., *4*, 91 (1970).
DeVries, K. L., J. Polymer Sci. C, No. 32, 325 (1971).
Kausch, H. H., J. Macromol. Sci. Chem., *C4*, (2), 243 (1970).

Kawashima, T., Nakamura, M., Shimada, S., Kashiwabara, H. & Sohma, J., Rept. Prog. Polymer. Phys. Japan, *12*, 469 (1969).

Moriuchi, S., Shimada, S., Nakamura, M., Kashiwabara, H. & Sohma, J., Polymer, *11*, 630 (1970).

Nara, S., Kashiwabara, H. & Sohma, J., J. Polymer Sci. A-2, *5*, 929 (1967).

Shimada, S., Kashiwabara, H. & Sohma, J., J. Polymer Sci., A-2, *8*, 1291 (1970).

Shimada, S., Kashiwabara, H. & Sohma, J., J. Polymer Sci., A-2, *6*, 1435 (1968).

Peterlin, A., J. Polymer Sci. C, No. 32, 297 (1971).

Discussion

Peterlin

I am extremely pleased by your data which so well corroborate the two component models of unoriented crystalline polymers with spherulitic structure. The amorphous layer between two parallel lamellae is occasionally bridged by a rather relaxed tie molecule originating from a macromolecule which started to be included independently in the crystal lattices of the adjacent lamellae so that the intermediate section is prevented from crystallization. Such an effect is typical for macromolecules of sufficiently high chain length and does not occur with low molecular weight material. Nitric acid treatment etches away the amorphous layer and also the relaxed tie molecules. Hence the mechanical separation of the sample during sawing which mainly proceeds through the amorphous layers between the lamellae breaks the tie molecules bridging these layers thus producing the radicals observed. No radicals are formed in the cases where no tie molecules are present.

Charlesby

Radicals are produced both in the crystalline and the amorphous regions by γ irradiation. But they are produced in fresh surface in the case of mechanical fracture. What do you think about the case of styrene which has no crystal regions?

Sohma

We have never tried mechanical fracture of polystyrene. However, if you have interest in the case of mechanical fracture of an amorphous polymer like polystyrene, I should like to tell you the case of *cis*-polybutadiene, which is also typical amorphous polymer. ESR spectrum was observed from the mechanical fractured polybutadiene at 77 K. This fact indicates that the chemical bonds were ruptured mechanically even for the amorphous polymer and the molecular motions of the polymers even in the amorphous part were so much hindered at 77 K as those in crystalline parts.

ESR Investigation of Chain Rupture in Strain Polymer Fibers

By A. Peterlin

Camille Dreyfus Laboratory, Research Triangle Institute, Research Triangle Park, N. C. 27709, USA

Summary

The number of ruptured chains in a strained fiber is much larger than the number of tie molecules in one cross section of the fiber and much smaller than the total number of tie molecules in the sample. One concludes that the rupture occurs only on discrete places with strain concentration caused by structure inhomogeneity of the fiber. According to the model of fibrous structure the ends of individual microfibrils are point vacancies of the microfibrillar superlattice when the longitudinal material connection by tie molecules is interrupted. Under applied stress the point vacancies open and form disk-shaped microcracks. Depending on the ratio of autoadhesion and strength of microfibrils the coalescence of microcracks can proceed axially along the boundary between microfibrils thus avoiding tie molecules and therefore rupturing very few chains or radially by breaking microfibrils with the great many tie molecules. The former case seems to apply to polyethylene, the latter to nylon.

Introduction

The ESR investigation of strained fibers shows the appearance of radicals well before the fracture occurs (Fig. 1) (Kausch-Blecken von Schmeling, 1970). Depending on the sensitivity of the instrument and the size of the sample one observes radicals already at about 40% of the strain to break. Their number depends on the strain and not on the stress. If one intercepts the straining at any strain value the radical population increases for a rather short while up to a limiting value characteristic for the strain and remains constant if by an inert atmosphere and low temperature one prevents radical decay. Lowering the strain or complete unloading do not affect the number. In a second run no new radicals are formed up to the maximum strain of the first run. The final rupture of the fiber does not produce any spectacular increase of the number of radicals.

The maximum number of radicals observed is particularly large in nylon-6

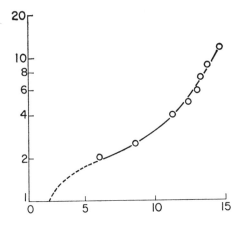

Fig. 1. *Abscissa:* % strain, ε; *ordinate:* number of radicals, $R \times 10^{-17}/cm^{-3}$.
Radical concentration (arbitrary units) in strained nylon 6,6 biaxially oriented ribbon (Verma & Peterlin, 1970).

and 66, about 5×10^{17} cm^{-3}, and substantially smaller in polyethylene and polyethylene terephthalate (Table 1). But even the smallest number is many times larger than that obtained if the chains were perfectly aligned parallel to the fiber axis and all of them ruptured in a single fracture plane through the sample i.e., $2/A_c = 10^{15}$ cm^{-2} in the case of polyethylene or nylon with the chain cross section $A_c \sim 20$ Å^2 under the assumption that one chain rupture produces two radicals. That means that the chain rupture must occur all over the sample and not only at the final crack (Campbell & Peterlin, 1968).

The radicals are assumed to be generated by polymer chain rupture. If the experiment is performed at room temperature the ESR spectra show that they are not primary radicals of the broken chain ends as formed at chain rupture but are secondary radicals on the unbroken chain formed by proton abstraction. In the case of nylon the primary and secondary radicals are

$$-CH_2-CO-NH-\dot{C}H_2$$

primary (chain end) radicals

$$\dot{C}H_2-CH_2-$$

$-CH_2-CO-NH-CH_2-\dot{C}H-CH_2-$ secondary (center of chain) radical.

Table 1. *Maximum Number of Radicals R Observed in cm³ of Fractured Samples, Activation Energy for Chain Fracture U_f (kcal mol⁻¹), Stress Coefficient γ and Chain Cross Section A_c in Crystal Lattice*

	R				
	Fiber	Film	U_f	γ	A_c
Nylon 6	5×10^{17} (3)	3×10^{16} (4)	45	160 Å³ (3)	19 Å²
Polyethylene	5×10^{16} (3)	5×10^{15} (4)	72		18
Polypropylene			56		35
Poly(ethylene terephthalate)	8×10^{15} (3)		38		

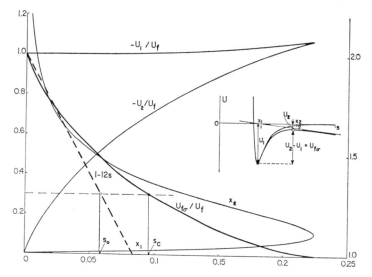

Fig. 2. *Abscissa:* $s = A\sigma r_0 / 12\, U_f$; *ordinate: (left)* U/U_f; *(right)* $x = r/r_0$.
Schematic representation of the activation energy change for bond rupture $U_{f\sigma} = U_2 - U_1$ under the influence of a tensile stress, σ. The straight line 1–12s represents the situation according to eq. (1) for $\delta = r_0$.

The formation of the secondary radical by the abstraction of a proton yields a saturated CH_3 group at the end of the originally broken chain. From the anisotropy of ESR signal (Verma & Peterlin, 1970) and the scavenging experiment with polymerizable monomers (Becht & Fischer, 1970) which can penetrate the amorphous but not the crystalline regions of the fiber structure one concluded that the radicals formed in strained fibrous material are located in the amorphous component which is less oriented than the crystal component. One generally agrees that during the fracture experiment one breaks the most strained tie molecules bridging the amorphous layer sandwiched between the crystal cores of consecutive lamellae (Peterlin, 1969).

The rupture of a strained polymer chain is a thermally activated process. According to Tobolsky & Eyring (1943) the average lifetime of a covalent chemical bond protected by an energy barrier U_f (activation energy for fracture) reads

$$t_\sigma = t_0 \exp{(U_f - \gamma\sigma)}/kT \tag{1}$$

where $t_0 = k/kT = 1.6 \times 10^{-13}$ s is the inverse of the natural oscillation frequency of chain groups, k is Planck's elementary quantum, σ is applied stress, and the stress coefficient $\gamma = A\delta$ is the activation volume of the bond. A is the area which concentrates the stress on one single chain and $\delta = r_2 - r_0 = r_0(x_2 - 1)$ is the bond extension up to rupture (Fig. 2).

The lifetime of nylon 6, viscose, and poly (methyl methacrylate) samples under dead load (Zurkov & Tomashevski, 1968) seems to be extremely well described by eq. (1) so that one may conclude that it reflects the direct role of the strained macromolecular chain. The sample fails as the chains rupture. Such a straightforward correspondence would preclude chain fracture before sample failure in contradiction with observations. A similar correlation between the stress and the fraction of ruptured chains is equally in disagreement with experiment, as can be best demonstrated by repeated straining close to the rupture point. In the first run the radical population is rapidly growing with strain so that one is tempted to derive the stress from the fraction of taut tie molecules which are just ready for rupture (DeVries, Lloyd & Williams, 1971). In the second run, however, the number of radicals observed remains constant, no new tie molecules are ruptured, but with exception of the shift of the stress strain curve as a consequence of the unrecoverable plastic deformation of the first run it remains practically identical with the virgin curve thus not showing any trace of correlation with chain rupture.

A more detailed examination of the energy field of the covalent bond shows that eq. (1) is a rough approximation valid for small σ only. If for simplicity sake, one assumes a Lennard-Jones type potential

$$U = -U_f(2x^{-6} - x^{-2}) \tag{2}$$

with $x = r/r_0 = 1 + \varepsilon$ and the equilibrium length r_0 of the bond the modification by the applied stress σ reads

$$U = -U_f(2x^{-6} - x^{-12}) - Ar_0\,\sigma(x - 1). \tag{3}$$

It displaces the minimum to x_1 and creates a maximum at x_2 as shown in the insert of Fig. 2. The values of x_1 and x_2 are plotted as function of the stress parameter $s = Ar_0\sigma/12U_f \sim \gamma\sigma/U_f$. As the minimum and maximum merge at $x_\infty = x_1 = x_2 = 1.1087$ the bonding ceases to exist. The applied stress reduces U_f to $U_{f\sigma} = U_2 - U_1$ plotted together with $-U_1$ and $-U_2$ in Fig. 2.

In the range yielding a lifetime of 1 h the ratio $U_{f\sigma}/U_f$ is between 0.4 and 0.7 if U_f/kT is between 100 and 50. In this range the $U_{f\sigma}$ calculated from the Lennard-Jones model deviates significantly from the linear relationship $U_f - \gamma\sigma$ of eq. (1). A substantially larger σ is needed for chain rupture than expected on the basis of the linear approximation. If one describes the phenomena by eq. (1) and finds a lifetime corresponding to a value $U_f - \gamma\sigma_0$ on the straight line which is the initial tangent of the $U_{f\sigma}$ curve the point on the U_f curve is situated at a far larger s_c value. The ratio s_c/s_0 is the factor by which γ derived from the linear relationship has to be multiplied in order to obtain the molecular parameter γ_c corresponding to chain rupture under the assumption of Lennard-Jones type potential energy of the covalent bond. Such a conclusion gets slightly modified but not basically altered by the choice of a

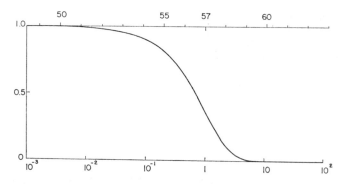

Fig. 3. *Abscissa: (bottom)* log t/t_o; *(top)* $\sigma_c/\text{kg mm}^{-2}$; *ordinate:* n/n_0.
Fraction of unbroken chains after a certain time *(lower scale)* according to eq. (4) or after 1 s as function of applied stress *(upper scale)* calculated from eq. (1) for $U_f = 45$ kcal mol^{-1} and $\gamma = 320$ Å^3 (nylon 6 according to Table 1 (Zhurkov & Tomashevsky, 1968).

different potential energy function which may be in better agreement with the quantum mechanics of chemical bonding.

The fraction of unbroken bonds at time t after loading turns out to be

$$n = n(t, \sigma) = n_0 \exp\left(-t/t_\sigma\right) \tag{4}$$

where n_0 is the number of unbroken bonds at $t=0$. The time dependence of radical population $R = 2(n_0 - n)$ shows up in the initial increase of ESR signal upon stress application. In a few minutes the increase becomes imperceptible.

The main consequence of eq. (4) is that there is no critical load above which all chains would be broken and below which all chains would remain intact. It depends on the time of the experiment, how large a fraction of chains under stress σ are broken, and observable by ESR of primary or secondary radicals. In Fig. 3 the fraction n/n_0 of unbroken chains is plotted over log t/t_σ. If one increases σ the decrease of t_σ is quite dramatic as seen from eq. (1). The reason for this high sensitivity of t_σ on σ is the high value of the factor γ/kT (Table 1) being 4×10^{-9} cm^2 dyn^{-1} for nylon 6. With $\sigma = 10$ kg mm$^{-2} = 10^9$ dyn cm^{-2} the product $\gamma\sigma/kT$ is already 4 yielding a factor $e^{-4} = 0.018$. Hence at fixed t the ratio t/t_σ and also log t/t_σ increases correspondingly fast. A small change in σ is sufficient for a move from the left side of the curve n/n_0 with almost all chains intact to the right side with all the chains ruptured. Therefore, for most practical purposes one can treat the strained chains as if they break instantaneously as soon as a critical strain or stress is reached, e.g.,

$$\sigma_c = (U_f + kT \ln t_0)/\gamma. \tag{5}$$

According to eqs. (1) and (4) at σ_c a fraction 0.368 of chains remains unbroken 1 s after load application and 0.007 after 5 s.

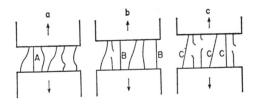

Fig. 4. Schematic view of the molecules bridging the amorphous layer and carrying the load up to their rupture. The average stress concentration in the three cases is $p/1$, $p/2$, $p/3$ where p is the number of chains in the crystal lattice.

Candidates for chain rupture are all chains oriented in the direction of the applied tensile stress, i.e., the chains in the almost perfectly oriented crystal lattice and the tie molecules bridging the amorphous layer between consecutive crystal lamellae. A simple consideration shows that the chances for the rupture of a chain in the crystal lattice are negligibly small. The area A concentrating the stress on such a chain is the chain cross section, about 20 Å² in the case of polyethylene or nylon. The bond extension up to rupture is about 0.3 Å yielding $\gamma = 6$ Å³ and for $U_f = 60$ kcal mol⁻¹ a critical stress 4×10^{11} dyn cm⁻² $= 4\,000$ kg mm⁻² which is between 50 and 100 times larger than the strength of a polymer solid. Even an increase of δ to 1 Å yielding $\gamma = 20$ Å³ and $\sigma_0 = 1\,200$ kg mm⁻² does not reduce σ_0 sufficiently and increase γ to the experimental value (160 Å³).

Tie molecules in the amorphous layers are more likely broken by the applied stress. The load is mainly carried by the most taut tie molecules (Peterlin, 1969) (Fig. 4). The average active area A concentrating the stress on a tie molecule equals A_c/β_t where β_t is the fraction of molecules in the crystal lattice which act as tie molecules in the amorphous layer. In the case of highly drawn polyethylene film between 10 and 30 % of molecules in the crystal lattice are tie molecules (Meinel & Peterlin, 1968). Such a factor β_t increases γ and reduces σ_0 to between 10 and 30 % of the value calculated for the chains in the crystal lattice. A further reduction takes place if one considers the fact that all tie molecules are not completely taut and consequently are broken in succession as schematically represented in Fig. 4. If the fraction of the most taut tie molecules is $\beta_{tt} < \beta_t$ one has $A = A_c/\beta_{tt} > A_t/\beta_t$. By such stress concentration the active area A and the stress coefficient can be so much increased that one obtains a reasonable agreement between observed bulk strength and the rupture of polymer chains. Hence in principle, the mechanism of chain rupture in a strained fiber seems to be compatible with general ideas about chain strength and fiber morphology, i.e., alternation of crystalline and amorphous layers in the fiber direction and the presence of a finite number of tie molecules in the amorphous layers connecting subsequent crystal lamellae.

Under the applied stress σ the sample is elongated to the macroscopic strain ε. In the alternating arrangement of crystals and amorphous layers the deformation of the crystals is negligible. Hence the strain of amorphous layer $\varepsilon_a = \varepsilon/(1-\alpha)$ is much higher than ε when the volume crystallinity α of the sample is above 0.50. With $\alpha = 0.70$ and $\varepsilon = 15\%$ one obtains $\varepsilon_a = 50\%$ which is quite a high strain for a taut tie molecule. According to Fig. 2 the bond is already stretched beyond the maximum stable value ε_m of 10.9% so that it cannot resist rupture if the force is kept constant. If, however, the strain is maintained or even increased and the force adjusted to the corresponding reaction force of the bond the lifetime of the bond is still quite remarkable although as a consequence of the large strain one is already so high on the potential energy curve that the difference $U_2 - U_1$ determining the lifetime is substantially reduced. And this is very likely what happens because with increasing strain more tie molecules get taut and share the carrying of the load so that the load concentration on each tie molecule is decreasing.

Such an effect makes the fracture of a microfibril more dependent on the total number of tie molecules in a single amorphous layer than on their tautness. If with further increasing load and strain no or not enough new tie molecules are mobilized for load sharing the fracture of all tie molecules in a cross section of the microfibril is inevitable and catastrophic as soon as the force contribution increasing with ε of not yet completely stretched tie molecules with ε_f below ε_m is not sufficient for compensating the force contribution decreasing with ε of those overstretched with ε_f above ε_m. Hence one is tempted to conclude that a microfibril is ruptured in a single stroke thus enlarging the microcrack. Such a process agrees very well with the model proposed by Bueche & Halpin (1964) for the crack propagation in amorphous polymers which assumes that small bundles of molecules are ruptured in successive steps.

The situation becomes less satisfactory if one considers the number of ruptured chains. In the case of nylon it is about 5 000 times higher than the probable number of tie molecules in one cross section and 200 times smaller than the number of tie molecules per cm³ (Table 2). That means that at fiber fracture one has ruptured only half a percent of all tie molecules in the nylon fiber. The simplest conclusion is that such rupture occurs on selected places distributed more or less randomly over the sample where as a consequence of

Table 2. *Chain Statistics in Nylon Fibers*

Number of chains in crystal lattice $= 1/A_c$	5×10^{14} cm^{-2}
Number of tie molecules per amorphous layer $= \beta/A_c$ $(\beta = 0.1)$	5×10^{13} cm^{-2}
Number of tie molecules/cm³ $= \beta/A_c L (L = 100\ \text{Å})$	5×10^{19} cm^{-3}
Number of ruptured chains $= R/2$	2.5×10^{17} cm^{-3}

structural inhomogeneity the strain or stress concentration on the taut tie molecules reaches higher than average values. Indeed Zhurkov et al. (1969) derive from the continuous small-angle X-ray scattering the existence of a very large number, between 10^{15} and 10^{17} cm^{-3}, of flat disc-shaped microcracks oriented perpendicular to the applied stress. It is easy to agree that chain rupture primarily occurs at these microcracks. But the views diverge about their origin.

According to the model of fibrous structure developed by Peterlin (1967, 1969, 1972) from deformational studies of single crystals, thin membranes, and bulk samples of polyethylene and polypropylene, the microcrack nuclei seem to be preformed in the material during drawing which transforms the original unoriented spherulitic or moderately oriented row nucleated sample into the highly oriented fiber. The basic element of the fibrous structure is the microfibril pulled out of the original folded chain lamella. It consists of fully oriented crystal blocks connected by a great many taut intrafibrillar tie molecules which originate from molecules partially unfolded during micro-necking. The lateral dimensions d of the blocks and hence of microfibrils are between 100 and 300 Å. The length $l_{mf}=k\lambda l_{ls}$ of the microfibril is proportional to the draw ratio λ and the length l_{ls} of the section of original lamella feeding the microneck producing the microfibril. The proportionality factor k is close to unity. In polyethylene one estimates the length l_{mf} to between 10 and 50 m roughly corresponding to a draw ratio between 5 and 25. That yields $n_v = 2/l_{mf}d^2 = 2 \times 10^{15} - 4 \times 10^{13}$ ends/cm^3.

At such point vacancies the axial material connection is interrupted and hence the stress and strain field locally distorted because very little force can be transmitted through them (Peterlin, 1971). Almost all the load, particularly at higher strain, has to be carried by the adjacent microfibrils because the vacancies can be easily opened by the applied stress so that physical holes are formed. The ensuing stress concentration $1/\beta_v \sim 1 + 1/m$ on the m adjacent microfibrils increases sufficiently the formerly mentioned factor $1/\beta_{tt}$ of the taut tie molecules for the rupture of the most affected microfibrils. By assuming 2×10^{15} point vacancies per cm^3 of nylon fiber with a draw ratio about 5 and one ruptured microfibril per vacancy one calculates the number of radicals formed per cm^3 to

$$R = 2n_v \beta d^2/A_o = 4\beta/A_o l_{mf} = 10^{18} \text{ cm}^{-3} \tag{6}$$

if $\beta = 0.1$ which seems to be a reasonable guess used already in Table 2. A straightforward analogy with polyethylene having β strictly proportional to the draw ratio up to $\lambda = 25$ with $\beta = 0.3$ would suggest $\beta = 0.06$ for $\lambda = 5$ and thus half of the value R from eq. (6). The orders of magnitude of the calculated and observed radical population are close enough for having confidence in the model.

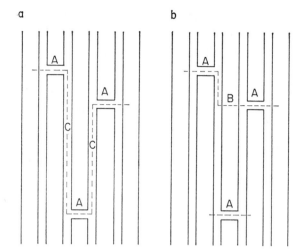

Fig. 5. Crack coalescence by (*a*) radial; (*b*) axial crack propagation. In the former case a microfibril is broken with all the tie molecules ruptured in at least one amorphous layer B. No or very few molecules (interfibrillar tie molecules) are ruptured in the latter case.

The first stage of microcrack formation at the point vacancies of the microfibrillar superlattice does not involve any substantial chain rupture because no or extremely few tie molecules are bridging the vacancy in axial direction. With increasing strain the microcracks tend to coalesce either by cracks through microfibrils perpendicular to the fiber axis or between microfibrils parallel to the fiber axis (Fig. 5). In the former case (radial crack propagation), new chains are broken and radicals formed and in the latter case (axial crack propagation) very few such ruptures occur. As soon as by such growth the critical size of crack is reached the sample fails catastrophically. But the number of ruptured chains at this last step—$\beta/A_c = 5 \times 10^{13}$ cm^{-2} with $\beta \sim 10\%$—is undetectably small compared to the total number, $10^{17} - 10^{18}$ cm^{-3}, ruptured in the great many microcracks before the catastrophic crack propagation begins.

Such a difference of microcrack propagation along the boundary between microfibrils and through the microfibrils is very likely the main reason for the difference in observed radicals between polyethylene and nylon 6 or 66. The strong hydrogen bridges in nylon very efficiently enhance the interfibrillar autoadhesion forces thus making the crack propagation along the boundary between adjacent microfibrils relatively difficult so that instead the microcracks prefer to grow by rupture of adjacent microfibrils. Such a rupture breaks the great many intrafibrillar tie molecules and hence yields a high radical population. In polyethylene, however, the autoadhesion forces between adjacent microfibrils seem to be smaller as demonstrated by the extremely small shear resistance of low density polyethylene drawn to a draw ratio

4.5. If that is indeed so, then the smaller interfibrillar forces favor crack propagation along the boundary between adjacent microfibrils instead of through the microfibrils thus breaking substantially less tie molecules.

The effect may be enhanced by the fact that the low draw ratio of nylon (~ 4) as compared with the substantially larger value (~ 15) of polyethylene produces more point vacancies and less intrafibrillar tie molecules. As a consequence, the number of point vacancies yielding microcracks is larger and the strength of the individual microfibril is smaller in nylon than in polyethylene. Both effects enhance the rupture of tie molecules in the former case and a crack propagation without significant chain rupture in the latter case.

The author would like to thank the Camille and Henry Dreyfus Foundation for generous support of this work.

References

See for instance the review article, Kausch-Blecken von Schmeling, H. H., J. Macromol. Sci., *C4*, 243 (1970).

Verma, G. S. P. & Peterlin, A., J. Macromol. Sci., *B4*, 589 (1970).

Zhurkov, S. N. & Tomashevski, E. E., Yield and Fracture (ed. A. C. Strickland) p. 200. Inst. Phys. & Phys. Soc. Conf., Ser. 1, Oxford 1968.

Zhurkov, S. N., Zakrevskii, V. A., Korsukov, V. E. & Kuksenko, V. S., Fiz. Tverd. Tela, *13*, 2004 (1971).

Campbell, D. & Peterlin, A., J. Polymer Sci., *B6*, 481 (1968).

Becht, J. & Fischer, H., Kolloid-Z. & Z. Polymere, *240*, 766 (1970).

Peterlin, A., J. Polymer Sci., A-2, *7*, 1151 (1969).

Tobolsky, A. & Eyring, H., J. Chem. Phys., *11*, 125 (1943).

DeVries, K. L., Lloyd, B. A. & Williams, M. L., J. Appl. Phys., *42*, 4644 (1971).

Meinel, G. & Peterlin, A., J. Polymer Sci., A-2, *6*, 587 (1968).

Bueche, F. & Halpin, J. C., J. Appl. Phys., *35*, 36 (1964).

Zhurkov, S. N., Kuksenko, V. S. & Slutsker, A. I., Fracture (ed. P. Pratt) p. 531. Chapman and Hall, London 1969.

Peterlin, A., J. Polymer Sci. C, *9*, 61 (1965); C, *15*, 427 (1967); C, *18*, 123 (1967).

Peterlin, A., Kolloid-Z. & Z. Polymere, *216/217*, 129 (1967); Man-Made Fibers, (ed. by Mark, Atlas & Cernia) chapter 8, pp. 283–320. Interscience-Wiley, New York 1967; Polymer Eng. Sci., *9*, 172 (1969).

Peterlin, A., J. Material Sci., *6*, 490 (1971); Text. Res. J., *42*, 20 (1972).

Peterlin, A., Intern. J. Fract. Mech., *7*, 496 (1971).

Discussion

Chapiro

The number of trapped radicals detected by ESR after fibre rupture should only be a small fraction (perhaps 10 %) of the total number of radicals produced, particularly since the radicals formed within a particular bundle must arise initially in a high local concentration.

Peterlin

I completely agree with your comment that the number of radicals observed is the lower limit for the number of ruptured chains. Molecular weight changes observed (Becht & Fischer, 1970) indeed show that either preferentially the longest chains are ruptured or a larger number of chains is ruptured than those derived from the number of radicals detected.

Hedvig

There is another kind of experimental evidence for the existence of voids or dislocations in stretched polyethylene: namely the observation of high electric field discharge patterns. The deformation of these patterns in the stretched polymer is in agreement with the model presented by Professor Peterlin. It is interesting, however, that in polypropylene the picture is not quite as clear as in polyethylene. In your Table 1 no figure is introduced for the radical concentration. Does this mean that in this polymer much less radicals are formed by mechanical treatment than the others?

Peterlin

Yes.

Sohma

According to your model the number of the broken chain increases with the increased strain. This means reduction in the number of tie molecules supporting the external force. Thus, it seems to me that the elastic constant must be decreased with the increased strain, or the increased strain may accelerate the break of the tie molecules because of the stress concentrated to the survived tie molecules. Is it right?

Peterlin

The fraction of broken tie molecules is at the maximum about 0.5%. The ensuing reduction of mechanical strength and elastic modules is undetectably small.

MOLECULAR ORBITAL CALCULATIONS

ESR Studies and MO Calculations of Radical Formation in DNA

By Thormod Henriksen

Department of Biophysics, Institute of Physics, University of Oslo, Blindern, Norway

Introduction

Among the large group of important biopolymers DNA seems to be the most outstanding since it was shown that this molecule contains the genetic information (Avery, MacLeod & McCarty, 1944). A large amount of research has been carried out in order to determine the structure and function of this macromolecule. It is well known that the Watson-Crick model of DNA as a double helix has been successful and has prompted ideas for replication as well as for the introduction of different modes of molecular changes which may have biological implications. It is obvious that all types of damage and/or molecular changes introduced to DNA are of general interest whether they result in mutations, cell death or only in minor reparable perturbations.

Different types of radiation such as UV, X rays and ionizing particles introduce molecular changes in DNA. Several experiments seem to indicate that DNA is the principal target with regard to lethal effects of radiation. Information about the types of damage caused in DNA by radiation is presently sought in experiments carried out along two different lines. Thus, in biological experiments with viruses and bacterial cells information down to the molecular level may be reached by changing the experimental conditions such as the oxygen content and by introducing radioprotective and radiosensitizing agents. On the other side, in physico-chemical experiments directly with DNA and its subunits it is possible to pin-point the type of damage caused in the more isolated molecular systems. It is of general interest if the results obtained in such experiments can be extrapolated to yield information of biological importance.

The molecular changes caused in DNA can be divided in two groups. Thus, in the first place, we have single and double strand breaks which may result in a major distortion of the molecule and the molecular morphology. In the second place, the DNA bases may be modified which in turn may lead to changes in the pool of information content and thereby in a mutation. Consequently, the molecular changes induced by radiation may have a significant effect on both replication and transcription.

The purpose of the present paper is mainly to demonstrate that ESR spectroscopy combined with molecular orbital (MO) calculations of spin density distributions in free radicals may yield information about the damage caused in DNA by radiation. It is necessary at this point to emphasize the fact that only unpaired spin products can be studied with the ESR technique. It is, however, generally accepted that these unstable intermediates are important products and they are involved in all molecular models proposed for the action of radiation on DNA.

History

The first attempts to study radiation induced radicals in DNA by the ESR technique were published more than 10 years ago. The results showed that powdered DNA usually yields one broad single ESR line which truly represents the composite pattern of the resonances from several different radicals. The attempts made to identify radicals based on the spectra of powder samples are relatively speculative.

Some type of a breakthrough came in 1963 and 1964 from two different sources. Thus, Ehrenberg A., Ehrenberg L. & Löfroth G. (1963) found that calf thymus DNA in an N_2 atmosphere yielded a spectrum quite similar to those obtained for the DNA components thymine and thymidine. And secondly, Pershan et al. (1964) found similar results for UV irradiated DNA. In experiments with deuterated thymine they presented quite good evidence that the radical in question was the 5-thymyl radical (I) in which a hydrogen atom is added to the C_6 position in the pyrimidine base thymine. It can be noted that the identification of this radical was possible even when a polycrystalline sample was used because only β protons with relative isotropic hyperfine interactions contribute to the observed pattern. This interpretation was subsequently confirmed in an experiment with a single crystal of thymidine (Pruden, Snipes & Gordy, 1965).

The early ESR experiments showed that large amounts of radicals were

(I)

formed and stabilized when DNA was irradiated. However, only one of the radical species was identified with certainty, even though the spectra showed that several other radicals were formed. Furthermore, the mechanisms for radical formation as well as the fates of the induced radicals were also mostly unknown. These problems have for a long time represented a challenge for the research work in this field.

Radical Identification

One of the first problems in physico-chemical studies of radiation events in DNA is to identify the different radical species. This is, however, an ambitious program due to the fact that radicals with different stability and reaction pathways are formed. Furthermore, some radicals disappear in combination reactions whereas others are transformed into more stable radicals, which implies that we must deal with both "primary" and "secondary" radicals as well as with intermediate reactions.

Radical identification by ESR spectroscopy is mainly based on a proper determination of the corresponding spin Hamiltonian which in a general case can be given as:

$$\mathcal{H} = \beta S \cdot g \cdot B + \sum_N S \cdot A_N \cdot I_N$$

where β is the Bohr magneton, S is the electron spin and I_N the nuclear spin operator and B the magnetic flux density. The spectroscopic splitting tensor is given by g and the hyperfine interaction tensor by A_N. The summation runs over all nuclei with non-zero spin which contribute to the observed spectra.

For identification purposes the spectroscopic splitting tensor g usually yields little information since all the ordinary hydrocarbon radicals exhibit a relative isotropic g tensor with only a small variation from one species to another. This implies that the hyperfine interaction term becomes the most important source of information for radical identification. Sufficient and necessary data for a proper determination of the different A_N tensors can only be found in single crystal experiments. Valuable information with determination of the isotropic values can be obtained in experiments on liquid samples where the rapid molecular movements in general will average out anisotropic terms. The smallest amount of information is obtained from powder samples or frozen solutions since the anisotropic terms contribute and result in blurred spectra. It can be noted that most experiments on DNA and its subunits have so far been made on powder samples. The identification of the 5-thymyl radical from such experiments is, however, more or less an exception based on the fact that the anisotropic terms are very small for this radical. Several other attempts to identify radicals from experiments along similar lines are, in the opinion of the present author, quite speculative.

During the last years we have worked on a program with the purpose to identify the induced radicals and to shed light on the mechanisms for radical formation and transformation in single crystals of DNA subunits. It is possible from observations on single crystals in three different planes to arrive at a determination of the different tensors included in the spin Hamiltonian. A proper radical identification is based on an application of the tensors data together with two other molecular properties:

(1) Primary radicals are usually formed in crystals either by an ionization event or by a rupture of a chemical bond whereby fragments appear. It is assumed that the largest radical fragment is formed from the intact molecule with only minor modifications in the structure. This implies that X ray diffraction data (if they exist) can be used in the identification procedure together with the directions of the anisotropic terms of the hyperfine interaction tensors.

(2) MO calculations of the electron spin density distribution in different radical candidates are used in order to arrive at the most plausible radiation induced product. This application of theoretical methods is possible since information about the unpaired spin density distribution gives rise to an estimation of the isotropic part of the A_N tensors which can be directly compared to the experimental values. This approach may give valuable support for one particular radical product and at the same time render other radical candidates unlikely.

Several papers have recently been published about the radicals induced in single crystals of DNA subunits. (Alexander & Gordy, 1967; Cook, Elliot & Wyard, 1967; Dertinger, 1967; Bernhard & Snipes, 1968; Herak, Galogaza & Duleic, 1969; Henriksen & Snipes, 1970; Hütterman, 1970). It is not the purpose of the present paper to review this large field, but rather to make an attempt to give an outline of the general approach by taking a couple of examples. For this purpose some recent experiments with the pyrimidine base thymine will be given. This DNA component seems to be quite important and is involved in the models proposed for the action of both UV and ionizing radiation on DNA. We have already discussed the formation of the 5-thymyl radical, but should probably point out that this species seems to be a secondary product. The details of the mechanisms for its formation in different systems (i.e. whether thymine is irradiated alone or within a more complex system) are not known. Altogether seven different radical products have been suggested for thymine.

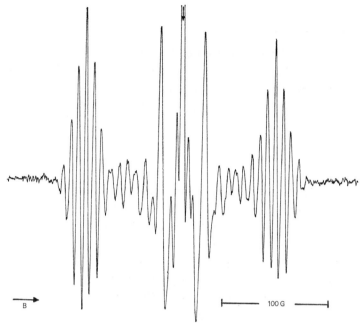

Fig. 1. Second derivative ESR spectrum of an irradiated single crystal of thymine anhydrate at 77 K. In this particular example the crystal is oriented with the external magnetic field almost along the crystallographic *c*-axis. The vertical arrow indicates the position of the resonance line for DPPH ($g = 2.0036$). An X band, reflection type spectrometer is used.

When single crystals of thymine are grown two modifications with and without "water of crystallization" appear. (The crystal data have been published by Gerdil, 1961 and Ozeki, Sakabe & Tanaka, 1969.) When an anhydrate crystal is irradiated at 77 K a very anisotropic ESR pattern is obtained. A characteristic example is shown in Fig. 1. It can be noted that the spectra extend over approx. 300 G, which is a much broader range than that usually found for organic radicals. Furthermore, when the crystal is rotated in the magnetic field a large anisotropy was observed which suggests that some type of a triplet state is involved. We would like to stress the fact that this large anisotropy would result in a broad unresolved resonance containing very little information if powder samples or frozen solutions were used. Only experiments with single crystals yield sufficient information for identification purposes.

The ESR spectra obtained by rotation in three planes and the subsequent annealing experiments showed that the resonance in Fig. 1 is composite consisting of the resonances from different radical species. Thus, one of the radical species gives rise to the well resolved groups of lines which can be seen on each side of the spectrum. In the following we shall briefly mention the main arguments which have led to the conclusion that a pair of radical cations are responsible for this resonance.

A Thymine Radical Pair

As can be seen from Fig. 1 the groups on each side of the central resonance consist of 7 well resolved lines with a splitting of approx. 7 G. This latter splitting was found to be quite isotropic, whereas the splitting between the two groups changes rapidly with orientation. For a radical pair the proper spin Hamiltonian is given by:

$$\mathcal{H} = \beta S \cdot g \cdot B + S \cdot D \cdot S + \sum_N S \cdot A_N \cdot I_N$$

where D is the dipolar tensor. The two unpaired electrons in the pair are separated by a distance (R) and are related by the formula $S = S_1 + S_2$. The system can be treated as if the spins were concentrated on each member of the pair and the dipolar tensor is therefore axially symmetric about the interspin vector (i.e. the direction of the line combining the two unpaired spins). A first order treatment yields the three energy levels ($S = 1$, and the values of M_s is therefore 1, 0 and -1). The two groups of ESR lines ($\Delta M_s = 1$ transitions) are found for the magnetic field values;

$$B = B_0 \pm \tfrac{1}{2} D$$

The axially symmetric tensor D is given by;

$$D = 3g\beta(3 \cos^2 \theta - 1)/R^3$$

where θ is the angle between the direction of the magnetic field and the interspin vector with length R. The values obtained in the experiment were; $D_{\parallel} = 180$ G; $D_{\perp} = -90$ G and the interspin distance $R = 6.76$ Å. The results implies that $D_{\parallel} = -2D_{\perp}$ which is in excellent agreement with that expected for $\theta = 0°$ and $90°$. The interspin distance is equal to the c axis and its direction in the crystal is a few degrees away from the c axis in the bc plane.

The next step in the identification procedure is to use the observed hyperfine interaction. It appears from the spin Hamiltonian that the energy level corresponding to $M_s = \pm 1$ are separated in the same way as that for the isolated radical. The energy level corresponding to $M_s = 0$ is, however, not splitt at all (the term $\sum_N S \cdot A_N \cdot I_N$ is zero) and since all $\Delta M_s = 1$ transitions involve this energy level the resulting hyperfine splitting will be only half that observed for the corresponding isolated radical.

The seven lines observed in Fig. 1 have an intensity distribution which can be given as $8:25:54:73:54:25:8$. This makes two interpretations possible. Thus, in the first place the intensity of the 5 middle lines corresponds quite well with that expected for 6 equivalent protons. However, the relative intensity of the two outermost lines (1:3 for the first two lines compared to the expected

1 : 7) makes the interpretation that only two methyl groups (6 protons) on adjacent molecules are involved relatively doubtful. An alternative interpretation is therefore that the groups consist of 9 lines with an intensity distribution similar to that expected for 8 equivalent protons (1 : 8 : 28 : 56 : 70 : 56 : 28 : 8 : 1). In this model it is reasonable to assume that the outermost lines are very weak and become masked within the noise level. As can be seen the intensity of the remaining 7 lines are in excellent agreement with that expected. It remains, however, to find the particular species (individual members of the pair) which fit into this scheme. This would for example require an isolated thymine radical with 4 interacting protons with almost identical splitting constants. The last step in the identification procedure is made possible by using MO-calculations.

MO Calculations

Molecular orbital calculations can, as mentioned above, be used to arrive at the distribution of the unpaired spin density within a radical. This information gives rise to an estimation of the isotropic hyperfine splitting constants. It is therefore possible to compare theoretically calculated splitting constants for different thymine radical candidates with those obtained by ESR spectroscopy. In our work we used the INDO method (Intermediate Neglect of Differential Overlap). For a more detailed description of this method see the work by Pople & Beveridge (1970). The INDO method for open shell calculations uses unrestricted determinantal wavefunctions which implies that different molecular orbitals are associated with different spin. In order to use the method the different radical candidates and the positional parameters for the atoms must be given with regard to a chosen reference system. A large program including a variety of radicals which may be induced in thymine is now under way in our laboratory. The positional parameters are varied slightly, but are based on the data obtained in X ray diffraction experiments (Gerdil, 1961; Ozeki, Sakabe & Tanaka, 1969). The results obtained so far show that the radical candidate which best can be associated with the results given above is the thymine cation. The calculated data (Fig. 2) yield an unpaired electron in $2p_\pi$ orbitals with the main spin density concentrated on N_1, C_4, C_5, and the two oxygen atoms. The main hyperfine splitting (the isotropic constants are given by the numbers in parentheses) is due to the proton bonded to N_1 and the methyl group protons. No rotation barrier was found for the methyl group and the three protons should be equivalent with a splitting constant of 13.1 G. It is of particular interest to note that this value is almost the same as that found for the N_1 proton which implies that altogether four protons in the thymine cation have the same splitting constant. Conse-

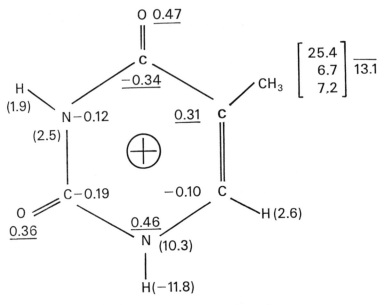

$E_T = -95.9230\,\mathrm{AU}$
$E_B = -\ \ 6.4853\,\mathrm{AU}$

Fig. 2. Some of the main results of the MO calculations for the thymine cation. The numbers in parentheses are the calculated isotropic hyperfine splitting constants given in gauss. For a rotating methyl group the average value for the three equivalent protons is 13.1 G. The negative value given for the N_1 proton is due to spin polarization and is of course not observed. The other numbers indicate α and β (minus sign) spin densities in the molecule. The largest values are underlined.

quently, a pair of radical cations has 8 equivalent protons which is in excellent agreement with observation. Furthermore, the calculated splitting constant of 12 to 13 G is in good agreement with that observed $(2 \times 7 = 14$ G). Whereas the calculated splitting constants for the other protons are too small to be observed, the value obtained for N_1 (10.3 G, which in the case of a radical pair should result in a splitting of about 5 G) is large enough to influence the spectra. The lack of nitrogen splitting in the observed ESR spectra makes the cation radical model somewhat uncertain. It should, however, be mentioned that our experience with MO calculations so far is that the INDO method sometimes yields a nitrogen splitting which deviates considerably from that observed.

In an attempt to sum up it can be concluded that irradiation of thymine at 77 K leads to the formation of several different radical species. The above discussion suggests that a pair of radical cations are formed by ionization of two nearby molecules. The radical pair formation is confirmed by the $\Delta M_s = 2$ transitions which yield a weak signal at $g \simeq 4$. With reference to the crystal data it can be concluded that the two thymine molecules involved in the pair are localized above each other in two different molecular planes along the *c* direction. The observed interspin distance of 6.76 Å clearly show that the planes involved are not nearest neighbours. This is also expected from the crystal structure which shows that the molecules are equivalent only for every second plane.

It was also shown that the radical pairs disappear significantly at temperatures above 150 K. The decay follows first order kinetics with an activation energy of 12.5 kcal mol^{-1} (details will be published elsewhere).

A Thymine Radical Anion

In order to demonstrate the combined use of MO calculations and ESR spectroscopy to attain information of the action of radiation on thymine, another example will be presented.

In a previous experiment (Henriksen & Snipes, 1970) we presented some ESR spectra for a single crystal of thymine monohydrate irradiated at 77 K. The spectra (examples are given in Fig. 3) were composite consisting of a weak signal from the 5-thymyl radical and a much stronger resonance which by us was characterized as a quartet and a doublet. The main reason for this division was that the relative intensities of the doublet lines and the quartet changed with the orientation of the crystal (see Fig. 3). The doublet was followed in all three planes (Fig. 3). The *g* value was almost isotropic and varied only from 2.0020 to 2.0030. The principal values of the doublet splitting were 15.2, 21.0 and 27.2 G suggesting that a fragment of the type $>\dot{C}$–H with α proton splitting was involved. The direction of the intermediate splitting constant was only 1° from the perpendicular of the thymine plane demonstrating that the electron was in a $2p_\pi$ orbital. The smallest splitting value was in a direction only 13° from the C_6–H bond direction indicating that the p_π orbital on C_6 is strongly involved. This prompted us to suggest the following neutral radical:

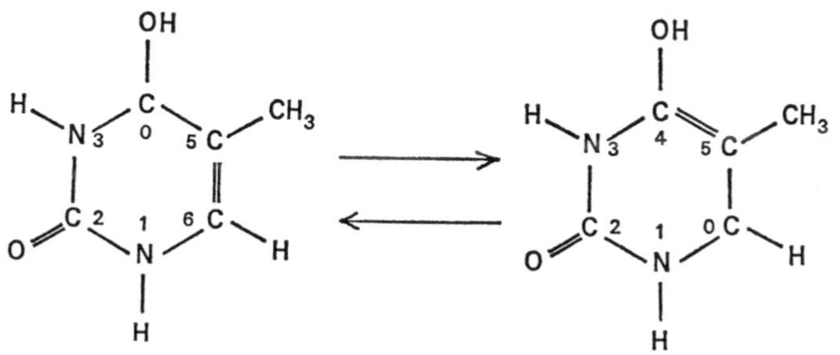

The radical is given by a resonance structure indicating that the main spin density is concentrated to p_π orbitals on C_4 and C_6. This would result in a

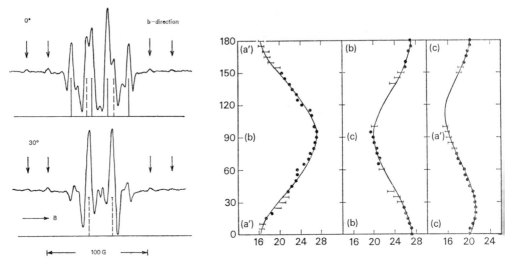

Fig. 3. *Abscissa:* doublet splitting (G); *ordinate:* angle/°.

ESR results for a single crystal of thymine monohydrate at 77 K. In the left part two examples of second derivative spectra are given. The crystal is oriented with the magnetic field in the *b* direction and 30° away from this axis in the *a'b* plane respectively. The vertical arrows indicate the positions of the two outermost lines on each side of the 5-thymyl radical spectrum. The bars give the positions of the quartet and doublet (dashed) resonance lines.

The splitting between the doublet lines in all three planes is given in the right part of the figure. The filled circles represent observations where the positions of the lines are well defined. The bars, on the other hand, refer to orientations in which the different resonances overlap and make the observations uncertain. The solid lines are calculated curves based on the tensor data.

hyperfine splitting characteristic for a $\rangle C_6$–H fragment as observed. The additional splitting which may be expected for the β proton in the $\rangle C_4$–O–H fragment is too small to be resolved since the proton presumably is in the plane of the thymine molecule yielding a large dihedral angle (θ) (it is known that β proton splitting depends on $\cos^2 \theta$). (For details see Henriksen & Snipes, 1970).

In our work we suggested that the radical may be formed by a proton transfer in the hydrogen bond between the oxygen on C_4 and the water of crystallization. This interpretation was supported by the observation that samples containing water of crystallization yielded far more radicals than the corresponding anhydrate samples. Furthermore, in thymidine (Snipes & Henriksen, 1970) which possesses a similar hydrogen bond at O_4 the same radical is found, whereas for dihydrothymine (Henriksen & Snipes, 1969) neither hydrogen bonding nor radical formation was observed. We suggested that the radical was formed by a proton transfer initiated by an electron capture. This mechanism would imply that a radical anion exists as an intermediate. It is therefore interesting to note that Sevilla (1971) in a recent work based on McLachlan SCF–MO calculations suggests that a radical anion

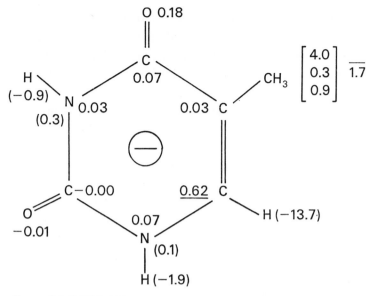

$E_T = -96.2479$ AU
$E_B = -\ 6.8102$ AU

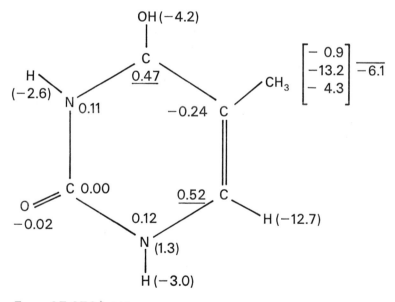

$E_T = -97.0724$ AU
$E_B = -\ 6.9660$ AU

Fig. 4. MO data for two of the proposed thymine radicals. Otherwise as for Fig. 2.

yields a doublet due to the C_6 proton. Sevilla's paper prompted us to carry out some INDO-calculations for both the radical anion and the radical suggested above (the main results are given in Fig. 4). It appears that both radicals can be characterized by a doublet splitting from the $\rangle C_6$–H fragment. For the anion radical the calculated splitting constants from the other nuclei are small which may imply that this interaction only results in broadening effects. Some fraction of the spin density is localized on the oxygen bonded to C_4 which may result in an increased g factor (due to the spin-orbit coupling of oxygen). No increase in the g factor from that for a pure hydrocarbon radical was found in the experiments discussed above.

The neutral radical suggested appears to be a typical π radical. The calculated splitting for the C_6 proton is smaller than that observed. Our experience with the INDO method is that the α proton splitting is largely influenced by the C–H bond distance. Fig. 4 shows that the splitting from a number of other nuclei, especially the methyl group, is relatively large. This raises the question whether some of this splitting could be observed even in a composite resonance pattern like that shown in Fig. 3. It should also be mentioned that the energy data indicate that the neutral radical is slightly more stable than the radical anion.

We would like to stop the discussion at this point even though a number of other examples from the DNA field could have been presented in order to demonstrate the type of information obtained by a combined use of ESR spectroscopy and MO calculations. In an attempt to sum up it can be concluded that the first example with the radical pairs clearly showed that identification was made possible only by a combination of single crystal work and MO calculations. In the other example the calculations made so far are, in our opinion, not sufficient to give a definite answer as to the most plausible of the two radical candidates. We are now in a situation where more information and experience about the reliability of the theoretical calculations of hyperfine splitting constants is desirable.

This brief review of our work on the action of radiation on DNA is based on experiments currently carried out by several members of our group. I would like to thank in particular R. Bergene and T. B. Melø for permission to use some of their results on radical pair formation in thymine prior to publication. Furthermore, the enthusiastic help from graduate students A. Heiberg, E. Sagstuen and G. Saxebøl is gratefully acknowledged.

References

Alexander, C. & Gordy, W., Proc. Natl. Acad. Sci. U.S., *58*, 1279 (1967).
Avery, O. T., MacLeod, C. M. & McCarty, M., J. Exptl. Med., *79*, 137 (1944).
Bernhard, W. & Snipes, W., Proc. Natl. Acad. Sci. U.S., *59*, 1038 (1968).
Cook, J. B., Elliot, J. P. & Wyard, S. J., Mol. Phys., *13*, 49 (1967).
Dertinger, H., Z. Naturforsch., *22*, 1266 (1967).
Ehrenberg, A., Ehrenberg, L. & Löfroth, G., Nature, *200*, 376 (1963).
Gerdil, R., Acta Cryst., 14, 333 (1961).

Henriksen, T. & Snipes, W., J. Chem. Phys., *52*, 1997 (1969).
Henriksen, T. & Snipes, W., Radiat. Res., *42*, 255 (1970).
Herak, J. N., Galogaza, V. & Duleic, A., Mol. Phys., *17*, 555 (1969).
Hütterman, J., Int. J. Radiat. Biol., *17*, 249 (1970).
Ozeki, K., Sakabe, N. & Tanaka, J., Acta Cryst., *B26*, 1038 (1969).
Pershan, P. S., Shulman, R. G., Wyluda, B. J. & Eisinger, J., Physics, *1*, 163 (1964).
Pople, J. A. & Beveridge, D. L., Approximate Molecular Orbital Theory. McGraw–Hill, New York 1970.
Pruden, B., Snipes, W. & Gordy, W., Proc. Natl. Acad. Sci. U.S., *53*, 917 (1965).
Sevilla, M. D., J. Phys. Chem., *75*, 626 (1971).
Snipes, W. & Henriksen, T., Int. J. Radiat. Biol., *17*, 367 (1970).

Discussion

Gillbro

You just found evidence for cation radicals in the irradiated thymine crystal. Would it not be possible to find anion radicals as well to account for the lack of negatively charged species?

Did you consider the possibility of a pair consisting of one anion and one cation radical? In that case one could expect altogether five coupling protons and one coupling nitrogen nucleus all with coupling constants of the same magnitude according to your MO calculations.

Henriksen

In a single crystal of thymine monohydrate, irradiated and observed at 77 K, we found a doublet which may be ascribed to a thymine anion. The identification is, however, difficult and not conclusive.

A radical pair consisting of one cation and one anion has been considered and rejected. This possibility appears less likely when the MO data are compared with the ESR results. Thus, the number of lines and the intensity distribution seem to be in closer agreement with the cation pair.

Williams

The hyperfine structure in your spectrum of the thymine radical cation is interpreted on the basis of four equivalent protons. Whereas the coupling to the three β protons from the methyl group is expected to be mainly isotropic, the coupling to the α proton on the nitrogen atom should be strongly anisotropic. Did you observe any breakdown in the equivalence of these four protons at certain orientations?

Henriksen

It should be noted that the two groups of lines can be observed outside the central resonance only within a relatively narrow angular region. Within this region the splitting appears quite isotropic.

P. Smith

You mention that your INDO-MO calculations sometimes give nitrogen splittings which are considerably at variance with experiment. I wonder if you can suggest any possible explanations for this phenomenon?

Henriksen

In order to arrive at the isotropic splitting constants from the spin density values you must include an empirical factor. This is a weak point in the calculations and the value used for nitrogen has been discussed already. I think we need more experience with the method and I hope it would be possible to arrive at a more close correspondence between the calculated and observed data.

Chapiro

I still find it difficult to accept that radical pairs or radical-ion pairs arise with such a high probability. I would expect pairs to form and remain stable only as quite exceptional events.

Perhaps the study of species formed by irradiation at very low temperature could give a clue to this problem.

Henriksen

I would like to make two comments. In the first place, the present experiments are carried out at 77 K. This implies that both irradiation and observation are made at that temperature. The radical pairs disappear already at temperatures above 130 K. Secondly, most of the resonance at 77 K is due to other radicals. At present the nature of these radicals is not known. I have in this paper used the radical pair to demonstrate our general method and how MO calculations may yield support to the ESR observations.

Kinell

Are the members of your radical cation pair oriented in the same way as the original molecules or have the members rotated in their planes due to changes in hydrogen bonding?

Henriksen

In these experiments the molecules have not moved or rotated. We can compare our results directly with the crystal data for the undamaged molecule. In fact, we used the crystal data extensively in order to arrive at the proposed model for the radical pairs.

D. R. Smith

In thymine many of the possible neutral radicals or the radical cation may be related to species formed from ammonia or aliphatic amines, e.g. $\dot{N}H_2$,

$\dot{N}H_3^+$, $R\dot{N}H$, RNH_2^+, $R\dot{C}HNH_2$, $R\dot{C}HNH_3^+$. The isotropic hyperfine coupling parameters of these species are well established and the anisotropic nitrogen couplings for NH_2 have been reported by J. R. Morton and myself. These data are conveniently summarized and referenced by Wardman P. & Smith, D. R., (Can. J. Chem. (1971)) in a paper on ESR studies of irradiated amines.

If one assumes that species formed from thymine and from amines have similar orbital hybridization, you could use these measured parameters plus your calculated spin densities to improve your predictions of ESR spectra. The data I refer to are all based on ^{15}N and deuterium substitution so they are quite reliable.

MOLECULAR MOTION IN POLYMERS

Spin Labeling Studies on the Dynamic State of Polymers

By P. Törmälä, H. Lättilä and J. J. Lindberg

Department of Wood and Polymer Chemistry, University of Helsinki, 00100 Helsinki 10, Finland

Introduction

In recent years the relation between the dynamic properties of the macro- and microstate of polymer melts and solutions have been the subject of numerous investigations. In this connection various magnetic resonance methods have been of special value for the investigations. Thus, in a series of papers our research group has studied these problems using the spin labeling technique and various polymers, especially polyethylene glycols (PEG), as models (Törmälä et al., 1970; Törmälä, Silvennoinen & Lindberg, 1971; Lindberg, Törmälä & Lähteenmäki, 1972). The present paper is a continuation of this work and is devoted to the discussion of the energy of activation of the motion of the polymer chain.

In aqueous (D_2O) solution the Larmor frequency dependence of the proton relaxation time indicates two motional mechanisms for PEG, whereas in the pure liquid state a broad undifferentiated relaxation typical of an amorphous polymer is noted (Preissing & Noack, 1971). From dielectric relaxation measurements (Porter & Boyd, 1971) it is further concluded that the dynamic behaviour of solid PEG is consistent with a model for a crystalline solid where the relaxation takes place in discrete amorphous regions similar to the melt.

From ESR spectra of PEG-nitroxyl radical complexes it is further learnt that there is a linear relationship between relaxation time and the dynamic volume (ηM_n) in quite a large range of molecular weights of the polymer (Lindberg, Törmälä & Lähteenmäki, 1972). However, as was also found for small molecules (Edelstein, Kwok & Maki, 1969), the effective hydrodynamic radius calculated from the ESR relaxation time measurements using the simple Stokes-Einstein equation is much smaller than the geometrical dimensions of the molecule (Törmälä, Silvennoinen & Lindberg, 1971). The above mentioned facts indicate clearly the complex nature of the effects studied and that they cannot be directly expressed in simple macroscopic terms.

The theory of the rotational relaxation processes in the region of rapid rotations ($\tau_c = 10^{-11} - 10^{-9}$ s) has been discussed and reviewed by several authors (cf e.g., Kivelson, 1960; Atkins, 1972). Accordingly in the case of nitroxyl radicals the width of each line in the spectrum is given by

$$W = a_1 + a_2 M + a_3 M^2 \tag{1}$$

where the a_i contains information about the magnetic interaction anisotropies and τ_c. From the linear and quadratic terms the following expressions for τ_i are obtained:

$$\tau_i^{(1)} = \left[\frac{W_1}{W_0} - \frac{W_{-1}}{W_0}\right]\left[-\frac{15\pi\sqrt{3}\,W_0}{8b\,\Delta\gamma\,H}\right] \tag{2}$$

$$\tau_c^{(2)} = \left[\frac{W_1}{W_0} + \frac{W_{-1}}{W_0} - 2\right]\left[\frac{4\pi\sqrt{3}\,W_0}{b^2}\right] \tag{3}$$

where W_i are the line widths of the hyperfine components and H the external magnetic field.

$$b = \frac{4\pi}{3}[T_{zz} - \tfrac{1}{2}(T_{xx} + T_{yy})] \tag{4}$$

and

$$\Delta\gamma = \frac{-|\beta|}{\hbar}[(g_{zz} - \tfrac{1}{2}(g_{xx} + g_{yy})] \tag{5}$$

are the anisotropies in the g tensor and hyperfine tensor, T, respectively. We used for them the values given earlier (Ohnishi, Boeyens & McConnell, 1966) for 4-maleimide-2,2,6,6-tetramethylpiperidine-1-oxyl. In principle $\tau_c^{(1)}$ and $\tau_c^{(2)}$ are equal, but the experimental values differed somewhat from each other at low temperatures. Therefore the mean values of $\tau_c^{(1)}$ and $\tau_c^{(2)}$ are given.

In the region of slow rotations ($\tau_c = 5 \times 10^{-9} - 10^{-7}$ s) theoretical studies are done correlating the relative positions of derivate peaks and τ_c (Kuznetsov et al., 1971; McCalley, Shimshick & McConnell, 1972). Accordingly the values of τ_c in this region were calculated by comparing the positions of high field peaks in the ESR spectra to the corresponding ones in the region of infinitely low and high viscosity (Shimshick & McConnell, 1972).

The activation energy E_a of the relaxation process may be determined from the temperature dependence of the correlation time τ_c by the relation:

$$d\ln(\tau_c)/d(1/T) = -E_a/R \tag{6}$$

Results

In order to elucidate the nature of the micro state of the PEG polymer in the melt and in the solid state we have determined the correlation time of the rotational relaxation and its temperature dependence for spin labelled PEG ($M_n = 4\,000$) (1) and compared it with the corresponding correlation time of the monomeric esterified nitroxyl radical (2) suspended in the polymer moiety.

The results of the measurements and calculations are given in Table 1 and Figs 1–3.

An inspection of Fig. 3 and Table 1 shows that there are three different regions of the activation process of the spin labelled polymer, the region I of the melt, $T > 340$ K, which is characterized by a very small energy of activation, 26.3 kJ mol^{-1}, a middle region II, 322 K $< T <$ 340 K with a considerably higher activation energy, 110.3 kJ mol^{-1}, and a third region III, $T < 322$ K, with an intermediate activation energy, 83.1 kJ mol^{-1}. Furthermore the strong retardation of the relaxation motion is observed at 286 K for the labelled polymer and at 266 K for the free radical in PEG ($\tau_c \geqslant 10^{-7}$ s).

It is further of interest to note that the visually observed melting point of the polymer is noted at 63°C. The same transition temperature is found from the relaxation of the free radical, whereas for the spin labelled polymer a certain delay effect, 67°C is noted. On the other hand a still more pronounced difference is found for the beginning of the very slow motion ($\tau_c \geqslant 10^{-7}$ s)

Table 1. *Activation Energy of ESR Relaxation E_a and the viscosimetric Activation Energy E_v for Spin labelled PEG and Nitroxyl Radical (2) dissolved in PEG*

Molecular weight of PEG, $M_n = 4\,000$; M.P. 336K (63°C)

| Temperature range | Labelled polymer | | Free radical |
	E_a kJ mol^{-1}	E_v kJ mol^{-1}	E_a kJ mol^{-1}
I > 340 K (67°C)	26.3	38.0	—
I > 337 K (64°C)	—	—	21.9
II 322–340 K	110.3	—	—
II 322–337 K	—	—	96.3
III < 322 K (49°C)	83.1	—	56.9
IV < 286 K (13°C)	—	—	—
IV < 266 K (−7°C)	—	—	—

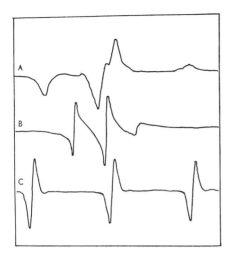

Fig. 1. ESR spectra of spin-labelled PEG 4 000 at different temperatures. A $T= -60°C$; $\tau_c = 10^{-7}$ s; B $T = 50°C$, $\tau_c = 2 \times 10^{-9}$ s, C $T = 150°C$; $\tau_c = 4 \times 10^{-11}$ s. A and B were measured at 100 G field; C at 40 G field.

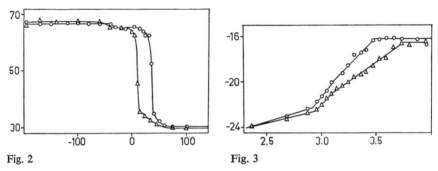

Fig. 2   Fig. 3

Fig. 2. *Abscissa:* $t/°C$; *ordinate:* B/G. A, $-\circ-$; B, $-\triangle-$.

The outermost peak-to-peak separation (G) as a function of temperature for spin labelled PEG (A) and for nitroxyl radical 2 dissolved in PEG (B).

Fig. 3. *Abscissa:* $10^3 T^{-1}/K^{-1}$; *ordinate:* $\ln \tau_c$. A, $-\circ-$; B, $-\triangle-$.

Ln τ_c plotted as a function of $1/T$ for spin labelled PEG (A) and for nitroxyl radical (2) dissolved in PEG (B).

i.e. 13°C for the spin labelled polymer against $-7°C$ for the free radical in PEG (Fig. 3). It may be concluded that the influence of the polymer chain as a whole on the segmental motion decreases very rapidly with increasing temperature. At the melting point there is only a very small difference left between the bulk viscosity and the internal viscosity of the polymer.

Discussion

A more quantitative description of the state of the transition region is obtained from the activation energy data in Table 1 and Fig. 3. The difference in activation energy between the labelled polymer and the free radical in region I is about 4.4 kJ mol^{-1} or a little more than one degree of rotational freedom,

0.5 *RT*. The activation energy of the relaxation of the spin labelled polymer is in addition 11.7 kJ mol^{-1} smaller than the corresponding activation energy of viscous flow, E_v at small shear rates, i.e. about *RT* in accordance with expectations owing to the differences in nature of the two processes. The finding of only one characteristic relaxation in the liquid state in PEG, which at decreasing the temperature changes abruptly into the corresponding relaxation of the stiff solid polymer (Figs 1, 2), indicates the homogeneous and random nature of the motion of the polymer chains in the melt. The results obtained by ESR measurements are so in accordance with NMR and dielectric studies.

On decreasing the temperature the difference between the relaxation activation energy of the polymer and the free radical increases rapidly to a value of 26.3 kJ mol^{-1} in range III corresponding to slower motions. This indicates clearly the locking action of the polymer chain.

The maximum in activation energy (E_a) of the spin labelled polymer in the region II, i.e. 110.3 kJ mol^{-1} as compared to 26.3 kJ mol^{-1} and 83.1 kJ mol^{-1} in regions I and III respectively, deserves special mention. It is probably not an artefact. It has been reported (White & Lovell, 1959) that the ordered conformation of PEG is preserved when going from the solid state to the melt. It is evident that a high E_a value of region II reflects the ordered nature (e.g. partial crystallization) of solid polymer.

The present findings seem to indicate some conformational changes, probably uncoiling, in the transition range during the melting process. A marked chain motion is certainly noted already about 50°C below the visually observed melting point. However, further investigations on the relation between the activation energy of the relaxation process and the molecular weight of PEG, which are under progress, will throw light on these questions.

Experimental

3-Chloroformyl-2,2,5,5-tetramethylpyrroline-1-oxyl was prepared (Krinitskaya, Buchachenko & Rozantsev, 1966) and coupled either to the commercial PEG ($M_n = 4\,000$) (*1*) or to methanol to give the nitroxyl radical ester (*2*). The substances were degassed and freed from oxygen very carefully in a high vacuum before use. The details of procedures will be given elsewhere (Törmälä, Lindberg & Lättilä 1972).

The ESR spectra were obtained from solid and melted samples on a Varian E-4 spectrometer operating at a microwave frequency of 9.1 GHz. The magnetic field was modulated with an amplitude less than one-sixth of the linewidth. The measurements were made in the range 120–425 K using a thermostat with an accuracy of $\pm 2°$. The radical concentration was 1:20 000 repeating units.

References

Atkins, P. W., Advan. Mol. Relaxation Processes, *2*, 121 (1972).

Edelstein, N., Kwok, A. & Maki, A. H., J. Chem. Phys., *41*, 179 (1969).

Kivelson, D., J. Chem. Phys., *33*, 1107 (1960).

Krinitskaya, L. A., Buchachenko, A. C. & Rozantsev, E. G., Zh. Organ. Khim., *2*, 1301 (1966).

Kuznetsov, A. N., Wasserman, A. M., Volkov, A. U. & Korst, N. N., Chem. Phys. Lett., *12*, 103 (1971).

Lindberg, J. J., Törmälä, P. & Lähteenmäki, M., Suomen Kemistilehti (1972). In press.

McCalley, R. C., Shimshick, E. J. & McConnell, H. M., Chem. Phys. Letters, *13*, 115 (1972).

Ohnishi, S., Boeyens, J. C. A. & McConnell, H. M., Proc. Natl. Acad. Sci. US, *56*, 809 (1966).

Porter, C. H. & Boyd, R. H., Macromolecules, *4*, 589 (1971).

Preissing, G. & Noack, F., Kolloid-Z. Z. Polymere, *247*, 811 (1971).

Shimshick, E. J. & McConnell, H. M., Biochem. Biophys. Res. Comm. *46*, 321 (1972).

Törmälä, P., Martinmaa, J., Silvennoinen, K. & Vaahtera, K., Acta Chem. Scand., *24*, 3066 (1970).

Törmälä, P., Silvennoinen, K. & Lindberg, J. J., Acta Chem. Scand., *25*, 2659 (1971).

Törmälä, P., Lindberg, J. J. & Lättilä, H., in preparation (1972).

White, H. F. & Lovell, L. M., J. Polymer Sci., *41*, 369 (1959).

Discussion

Sohma

In your text you mentioned "the results obtained by ESR are in accordance with NMR and dielectric studies". What do you mean "be in accordance"? ESR results are not necessarily in accordance with NMR and dielectric measurements with regard to the activation energy as well as the correlation time, because the time constant associated with ESR, i.e. the microwave region, is quite different from those of either NMR or dielectric constant and also because the correlation of ESR may be different from those of NMR and dielectric properties.

Lindberg

The general conclusions drawn from spin labelling experiments are in accordance with dielectric and NMR measurements. The numerical data are naturally different.

Hedvig

I think that one should not expect good agreement between the results obtained by NMR and by spin labelling ESR technique on one hand and those of the dielectric and mechanical relaxation on the other, because the correlation functions are different.

By the way I believe that one should always use the classical relaxation methods besides NMR and ESR. The introduction of the labelling material into the polymer might change the structure significantly.

Lindberg

The concentration of labelling groups is very small $\approx 1:20\ 000$.

P. Smith

I was impressed that you were able to label PEG samples of M_n equal to 4 000 at one of the two terminal OH groups. I would appreciate learning whether your attempts to do so encountered procedural difficulties and, if so, how these were overcome.

Also, I would be interested in knowing how the hydrodynamic-radius values you determined *via* ESR measurements compared quantitatively with the hydro-dynamic-radius values which you were able to calculate, with the use of non-ESR data; and any comments about the differences between these two sets of values.

Lindberg

The spin labels are naturally added statistically and the addition of one nitroxyl group to one end is naturally more probable than the addition to both ends. The difference in hydrodynamic radius obtained from viscosity and label-ling data has been discussed frequently. Today no clear explanation has been found.

Tsuji

You got activation energies for molecular motion from one kind of labels. Is there any possibility that the values depend on the kind of labels? Larger labels are expected to move more slowly.

Lindberg

It is possible.

Molecular Motions in Solid High Polymers Reflected in ESR Studies

By *Hisatsugu Kashiwabara, Shigetaka Shimada and Junkichi Sohma*

Nagoya Institute of Technology, Showa-ku, Nagoya, Japan and Hokkaido University, Sapporo, Japan

Introduction

The ESR method is, intrinsically, to detect the electronic state of the local site near the unpaired electron. The method was applied in the field of polymer research for identification of the free radicals produced by irradiation at the earliest stage. In the course of the studies with this aim, we experienced many difficulties in making the clear identifications of the free radicals trapped caused by the very diffuse spectra observed in many cases. Causes of the difficulties can be considered to be the following.

(i) Irradiation is made for trapping of the free radicals, usually, and the action of the ionizing radiation is very complicated, one resulting in many kinds of free radicals trapped. The ESR spectrum observed is, therefore, a superposition of the many kinds of spectra corresponding to the respective free radicals.

(ii) Various physical states in high polymer substances behave differently in trapping free radicals and give the different circumstances for hyperfine interactions. Variations of the physical states appear in the variations of the steric configurations near the unpaired electrons and the spectra for free radicals with different steric configurations must have the different hyperfine splittings, though the chemical structures of them are the same. Therefore the resultant spectrum observed must be a superposition of the different kinds of spectra. This is the reason why the spectrum observed for the polymer radical is very complicated.

According to the situations described above, identification of the free radicals trapped in polymers was not an easy task. However, several devices removing the difficulties were successful for some polymers and the free radicals trapped in them were identified rather clearly. Cases for polyethylene, (Kiselev et al., 1960; Lawton et al., 1960; Ohnishi et al., 1961; Kashiwabara,

1961), polypropylene (Yoshida & Rånby, 1965; Fischer, Hellwege & Johnsen, 1960), polytetrafluoroethylene (Rexroad & Gordy, 1959; Tamura, 1962) and polymethylmethacrylate (Abraham et al., 1958; Ingram, Symons & Townsend, 1958; Sohma, Komatsu & Kashiwabara, 1965) are the well known examples of the successful studies. One of the devices, which remove the difficulties in identifications of the free radicals, was the separation of the superposing spectra by the vanishing of the fast decaying radical species. This procedure needs, as the simplest way, heating of the sample.

For this reason, heat treatment of the samples trapping free radicals produced by, for example, irradiation at lower temperature, was found to be an indispensable technique in the ESR studies of solid high polymer. In the course of the studies including the heat treatment, we had interesting experiences when finding that the decays of free radicals trapped in solid high polymer reflected the characteristic features of the matrix polymers. On pp. 276–281 we will report on studies of the decay of free radicals and its relation to the molecular motion in matrix polymer. On pp. 281–285 an interesting phenomenon of free radical migration and its correlation with the decays of free radicals will be discussed.

Decay Curves

Definition of the Decay Curve

After the evacuation and the γ irradiation at liquid nitrogen temperature of the materials sealed in the ampoule, ESR spectrum was observed at liquid nitrogen temperature. (Liquid nitrogen temperature will hence be abbreviated LN). Then, the sample was heated at a higher temperature than that of liquid nitrogen, say, the temperature of liquid oxygen, for 5 min, and was cooled again to LN. The ESR spectrum was observed at LN. These procedures were repeated successively, each at much higher temperatures below the melting point. Since the heat treatment causes partial decay of the free radicals trapped, the spectral intensity obtained after each of the heat treatments can be reduced. Ratios of the spectral intensities obtained after the heatings to that for non-heat treatment were plotted against the temperatures of the heatings. This plot will be called the decay curve. Decay curves obtained for several polyethylenes are shown in Fig. 1.

Results and Discussions for the Case of Polyethylene

Decay curves of polyethylene: Fig. 1 (Kashiwabara, 1964; Nara et al., 1968) shows the decay curves of several polyethylenes. S-1 means HDPE (Sholex $-6\,000$) annealed at 165°C for 5 h, recrystallized at 110°C for 24 h and then cooled

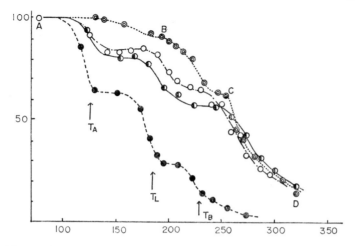

Fig. 1. *Abscissa:* T/K; *ordinate:* relative intensity ($\%$).
Decay curves of polyethylene: $S-1$ (○), $S-2$ (◖); $S-3$ (○); $M-1$ (●).

to room temperature very slowly. S-2 is the same material annealed at 170°C
for 43 h and then cooled very slowly from 170°C to room temperature. S-3
means the same material annealed at 170°C for 45 h and cooled quickly to
LN. M-1 means LDPE (Sumikathene-G-20) annealed at 160°C for 5 h, recrystallized at 110°C for 24 h, and then cooled very slowly to room temperature. Crystallinities of these samples are recorded in Table 1. Each of
the decay curves in Fig. 1 shows remarkable drops in three temperature
regions, which we shall designate as T_A, T_L, and T_B. Decay reactions at
several temperatures ranged in each of the temperature regions were also
studied. These were found to be of second order for all decay reactions
studied and activation energies of the decays in the respective temperature
regions were obtained from the knowledge of the temperature dependence
of rate constants. Table 2 shows the activation energies obtained. The
major part of the free radicals studied were the alkyl type radicals

$$\sim CH_2 - \dot{C}H - CH_2 \sim.$$

Decay reaction in the region T_L: The decay reaction in the region T_L has
some interesting features. The extent of decay in this region was plotted
against the degree of crystallinity of HDPE. In addition to the above
mentioned samples, S-1, S-2, S-3, several samples crystallized from dilute

Table 1. *Crystallinities*

Sample	S-1	S-2	S-3	M-1
Crystallinity ($\%$)	89	84	68	51

Table 2. *Activation Energies of Decay Reactions in PE*

Decay region	E (kcal mol^{-1})	
	HD (S-2)	LD (M-1)
T_A	0.4	0.7
T_L	9.4	23.1
T_B	18.4	24.8

xylene solutions were investigated for this plot. It was found that the decay in T_L for HDPE is proportional to the crystallinity. Table 2 reveals that the activation energy of the decay in T_L for HDPE is 9.5 kcal mol^{-1} and that for LDPE is 23.1 kcal mol^{-1}. On the other hand, recent studies by Hideshima et al. (1968) of γ' dispersion in polyethylene shows that two kinds of molecular motion can be associated with γ dispersion. They are the motion of molecules in lamellar surfaces and the local mode motions in the amorphous phase. Activation energy for the former was found to be 12 kcal mol^{-1} and that for the latter was 25 kcal mol^{-1}. The similar situation of the mechanical dispersion was also found for the case of polyoxymethylene. If this is the case for our results, the decay reaction in T_L for HDPE must be a reflection of the molecular motion associated with the γ dispersion in lamellar surfaces, and that for LDPE corresponds to the γ dispersion in the amorphous phase.

On the decay in T_A: Table 2 indicates that activation energies for the decay in T_A are very small both for HDPE and LDPE, and no corresponding dispersion region was found in mechanical studies. Amounts of decay of free radicals in T_A were found to be proportional to the amount of amorphous portions (Fig. 2). Therefore, the decays in T_A can be considered to be closely related to a very slight motion of molecules in the amorphous phase.

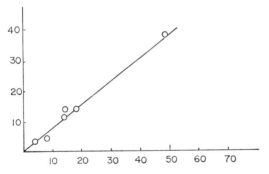

Fig. 2. *Abscissa:* content of amorphous part (%); *ordinate:* amount of decay in region T_A Relation between amount of decay in T_A and crystallinity.

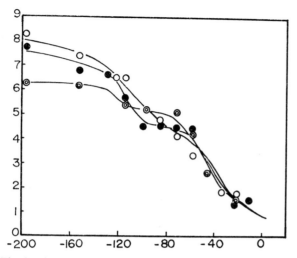

Fig. 3. *Abscissa: t/°C; ordinate:* radical conc. (spins/g) × 10⁻¹⁹.

Effect of mechanical fatigue to decay in T_A: original sample (◎); sample subjected to mechanical fatigue (●, ○). From Nagamura et al. (1971).

Free radical decay can be a very sensitive indicator of the motion of the matrix chain and it is possible to say that the decay of free radicals in T_A is a reflection of a very slight motion in the amorphous phase though the mechanical studies do not detect the similar phenomenon. Recently, Nagamura et al. (1971) reported that the extent of the decay in T_A was enhanced for the polyethylene subjected to the mechanical fatigue as shown in Fig. 3. It can be said that mechanical fatigue produced the movable sites of very slight motions discussed in the present section, and the result of Nagamura et al. (1971) seems to be consistent with our considerations on the decay in T_A. In connection with the above consideration, it must be noted that Hoffman et al. (1966) reported on small scale molecular motions with an activation energy less than 1 kcal mol⁻¹ for some polymers.

On the decay in T_B: As shown in Table 2, activation energies for decay processes both in HDPE and in LDPE are rather large and the temperature regions are very close to that of β dispersion in polyethylene. From these facts, we can say that the decay of the free radicals in T_B is a reflection of the molecular motion associated with β dispersion.

Decay Curves of Polypropylene and Other Polymers

Decay curve of isotactic polypropylene (Nara, Kashiwabara & Sohma, 1967) is shown in Fig. 4, which indicates two remarkable drops, near 170 K and 260 K, respectively. These decay regions will be designated T_A and T_B. In this study, isotactic polypropylene was annealed at 220°C for 40 h and cooled very slowly to room temperature. The major part of the free radicals studied for this

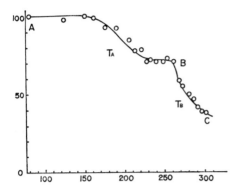

Fig. 4. *Abscissa:* T/K; *ordinate:* rel. intensity (%).
Decay curve of isotactic polypropylene.

decay process was the alkyl type radical, $\sim CH_2-\dot{C}(CH_3)-CH\sim$. Activation energies for the decays in T_A and T_B were found to be 11 kcal mol^{-1} and 48 kcal mol^{-1}, respectively. These values are very close to those obtained in mechanical studies; i.e., 13 kcal mol^{-1} for γ dispersion and 58 kcal mol^{-1} for β dispersion. The respective temperature regions of the decays are also very close to those for mechanical dispersions. These facts can be the convincing evidences that the decay processes of the free radicals are controlled by molecular motions in the solid polymer matrix.

Decay curves of the free radicals trapped in γ-irradiated polybutadiene (Tsuchihashi et al., 1969) were also obtained and it was found that the temperature regions of remarkable decays of the free radicals were corresponding to the temperature regions at which narrowing of the line widths of broadline NMR occurred for both kinds of polybutadienes. Fig. 5 shows the results and arrows mean the temperatures of narrowing of line widths in the broad line NMR studies.

Variation of the Steric Configurations of the Free Radicals

In addition to the correspondence between decay regions and mechanical dispersions discussed in the preceding sections, the steric configuration of

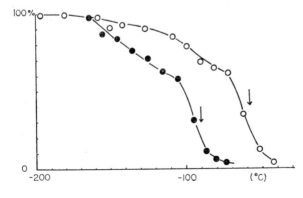

Fig. 5. *Abscissa:* $t/°C$; *ordinate:* relative intensity (%). Decay curves of polybutadiene: (○) *trans,* 70%; *cis,* 10%; 1,2-vinyl, 20%; (●) *trans,* 54.3%; *cis,* 34.3%, 1,2-vinyl, 11.2%.

the free radicals which decayed out in each of the decay regions, can be studied. For example, subtraction of the spectrum observed for point B in Fig. 1 from that of A makes the ESR spectrum of the free radicals which vanished during heating in the temperature range T_A. Similar procedures can be made for T_L and T_B of polyethylene. This kind of study (Nara et al., 1968) reveals that the free radicals which decayed in the lower temperature region, have a less uniform steric configuration than that for free radicals which decayed out in higher temperature regions. Similar studies were also made for polypropylene (Nara, Kashiwabara & Sohma, 1967).

Migration of Free Radical

Need of Migration Reaction of Free Radicals

On p. 280 it was discussed that decays of the free radicals were controlled by the molecular motions of the matrix polymers trapping the free radicals. On the other hand, however, most of the materials studied were of rather high crystallinities and the major part of the free radicals were considered to be trapped in the crystalline part at the earlier stage of the trapping process due to the irradiation. Therefore, it seems very peculiar that most free radicals in the crystalline part decay due to the motion of molecules in the amorphous part. In connection with this point, migration of the free radicals from the crystalline part to the amorphous part has necessarily to be considered. Recently, we observed a phenomenon, which can be considered to be a "direct" observation of the free radical migration along the chain of the polymer molecules in the case of polyethylene. This was closely related with our studies on photoinduced changes of the free radicals in polyethylene. A few years ago, we found a characteristic photoinduced phenomenon in the γ-irradiated polyethylene (Shimada, Kashiwabara & Sohma, 1970a); i.e., the polyethylene containing trapped allylic free radicals was subject to the main chain scission by illumination of light at a wavelength longer than 3 900 Å. This was considered to be a reflection of the fact that one of two kinds of excited states (Longuett-Higgins & Pople, 1955) of the allylic free radicals caused the main-chain scission reaction as shown in the following scheme.

$$—CH{=}CH—\dot{C}H—CH_2— \quad \rightsquigarrow \rightarrow$$
$$\sim CH{=}CH—CH{=}CH_2 + {}^{\cdot}CH_2—CH_2—$$
$$\qquad\qquad \rightarrow CH_3—\dot{C}H—CH_2—CH_2— \qquad (1)$$

Observation of this phenomenon was confirmed by the ESR method, UV-absorption spectrometry, and measurements of molecular weight. The ESR spectrum corresponding to allylic free radical was converted to the spectrum for the free radical $CH_3{-}\dot{C}H{-}CH_2{-}$ after illumination of light at longer wavelength. UV-absorption spectrometry indicated the increase of the diene structure and molecular weight after the illumination was found to be reduced. These results supported the possibility of the process stated in (1). The ESR spectrum corresponding to the free radicals, $CH_3-\dot{C}H-CH_2-$, appeared as an octet spectrum (Fig. 6a), and this identification was also confirmed by much more rigorous analyses in the case of solution grown polyethylene (Shimada, Kashiwabara & Sohma, 1969).

Continued studies of this phenomenon will involve the migration reaction.

Migration of Free Radicals in High Density Polyethylene

High density polyethylene like S-1 on p. 276–277 (Shimada, Kashiwabara & Sohma, 1970b) was studied and the following facts were found: the octet spectrum of Fig. 6a changed into a sextet spectrum (Fig. 6b) corresponding to an alkyl radical by heating at lower temperature (around -34 to $0°C$) for a shorter time period; the sextet spectrum changed into a septet spectrum corresponding to the allylic radicals again by heating at higher temperature than room temperature for a longer time duration. The former reaction will be called process (A); the latter process (B). The process (B) is the same phenomenon as reported by Ohnishi et al. (1963)

$$CH_3-\dot{C}H-CH_2-CH_2 \sim \xrightarrow{A}$$

$$CH_3-CH_2-\dot{C}H-CH_2 \quad \text{---} \quad CH_2-CH=CH_2- \qquad (2A)$$

$$\left. B \right\downarrow$$

$$\longrightarrow CH_3-CH_2-CH_2 \quad \text{---} \quad CH_2-\dot{C}H-CH=CH-CH_2- \quad (2B)$$

Process (A) was investigated at various temperatures ranged from -43 to $-20°C$; the changes of the parameter associated with intensities of the octet spectrum, a_8, were observed for various time durations of heating at $-43°$, $-35°$, $-28°$ and $-20°C$, respectively, and $-\log a_8/a_8^\circ$ was plotted against the time of the heating, where a_8° denotes the initial value of a_8. The plotted points were found to be linear and this means that the process (A) is of first order reaction. The meanings of the parameters, a_8 or a_8°, can be understood in Fig. 6. In Fig. 6, I_8 means the intensity of a characteristic peak of the octet spectrum and I_6 means that of the sextet spectrum. Fig. 6b is a spectrum observed after the conversion was completed. By the simple consideration about the situation that the observed spectrum at a stage of the process (Fig. 6c) is a superposition of the octet spectrum and the converted sextet spectrum, effective intensity of the octet spectrum can be proportional to a_8 given in the following equation.

$$a_8 = \frac{I_8 + I_6^\circ}{I_8^\circ + I_6^\circ} \quad (3) \qquad\qquad I_8 = a_8 I_8^\circ - (1 - a_8) I_6^\circ \quad (3')$$

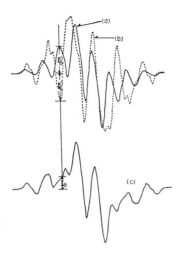

Fig. 6. ESR spectra of polyethylene.

From each slope of the linear plots of $-\log a_8/a_8^\circ$ against time, the rate constant, k, of the reaction at each temperature can be determined. Logarithms of k's for various temperatures were plotted against the inverse temperature and slope of the linear plots gives us the activation energy of the process. Plots in Fig. 7 (right) indicate the results. We have several linear plots in Fig. 7, and these correspond to the various irradiation doses; in other words, the experiments were made for the samples which were subjected to various irradiation doses in order to produce allylic radicals. Activation energies for various cases were found to be almost the same values, which is estimated as 18 kcal mol^{-1}. Similar experiments were also made for process (B) and the results are shown by the plots in Fig. 7 (left). Activation energy for this case was also obtained at around 18 kcal mol^{-1}. In Table 3 the results of the activation energy measurements are tabulated. And also, it must be noted that the absolute rate of process (A) is about 2 000 times that of process

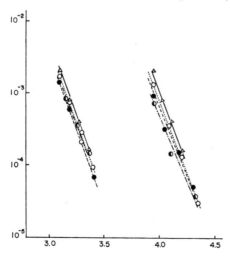

Fig. 7. *Abscissa:* $10^3\,T^{-1}/\mathrm{K}^{-1}$; *ordinate:* rate constant.

Arrhenius plots of rate constants: (right) octet → sextet; (left) alkyl → allyl. Dose: 3 (△—△); 20 (...○...); 50 (–·–○–·–); 100 (– –●– –) Mrad.

Table 3. *Activation Energies of Migration Reactions*

Dose (Mrad)	E (kcal mol^{-1})	
	Process A	Process B
3	19.2	18.8
20	18.8	18.6
50	17.1	16.1
100	18.6	19.5

(B). From these results, it is quite reasonable to say that both processes (A) and (B) can be caused by the same mechanisms of proton migration along the chain of polyethylene.

The k values were found to be slightly dependent on the irradiation dose as shown in Fig. 8. This seems to be a trace of migration along the chain also, because cross-linking caused by the irradiation disturbs the migration reaction and the increase of the cross-linking density causes the reduction of the migration reaction rate for both processes (A) and (B) as shown in the same figure.

A model of the migration process can be considered as follows: the free radical, for example, of alkyl type is the localization of the unpaired electron at carbon atom where the electronic configuration of sp^2 is expected and, therefore, the site with the electronic configuration of sp^2 can be expected to have a kind of distorted conformation of the backbone chain molecule from the usual planar-zigzag conformation. This distortion must behave so as to have a much higher potential than the other usual site. Thermal motion of the molecule can relax this kind of distortion; therefore the migration reaction process can be enhanced with raising of the temperature.

Though we have experienced several complicated situations concerning the processes in the other materials (Shimada, Kashiwabara & Sohma, 1971) observations of the processes (A) and (B) have been briefly stated as an evidence of the migration reaction and also as a reasonable way for inter-

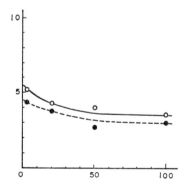

Fig. 8. *Abscissa:* dose/Mrad; *ordinate:* rate constant.

Relation between rate constant and dose: process (A) at $-31°$C (--●--), process (B) at 35°V (–○–).

preting the decaying features of the free radicals described in Part A. We believe that these results are the beginning of a new chapter, the reporting of which must be deferred until the completion of the work now in progress.

Mr S. Nara and Mr N. Tsuchihashi made some contributions in obtaining the results in this article. Fig. 3 was included in this article with the permission of Professor Kusumoto.

References

Kiselev, A. G., Mokulskii, A. & Lazurkin, Yu. S., Vuisoc. Soed., *2*, 1678 (1960).
Lawton, E. J., Balwit, J. S. & Powell, R. S., J. Chem. Phys., *33*, 395 (1960).
Ohnishi, S., Ikeda, Y., Kashiwagi, M. & Nitta, I., Polymer, *2*, 119 (1961).
Kashiwabara, H., J. Phys. Soc. Japan, *16*, 2494 (1961).
Yoshida, H. & Rånby, B., Acta Chem. Scand., *19*, 72 (1965).
Fischer, H., Hellwege, K.-H. & Johnsen, U., Kolloid Z., *170*, 61 (1960).
Rexroad, H. N. & Gordy, W., J. Chem. Phys., *30*, 399 (1959).
Tamura, N., J. Chem. Phys., *37*, 479 (1962).
Abraham, R. J., Melville, H. W., Ovenall, D. W. & Wiffen, D. H., Trans. Faraday Soc., *54*, 1133 (1958).
Ingram, D. J. E., Symons, M. C. R. & Townsend, M. G., Trans. Faraday Soc., *54*, 409 (1958).
Sohma, J., Komatsu, T. & Kashiwabara, H., Polymer Letters, *3*, 287 (1965).

The first four references cited above are selected from the papers in the earlier stages of the identifications of the polymer radicals. Therefore, these include only part of many papers in the field. I know we have many valuable papers which should be recorded in the history of ESR study of polymer. If papers of authors participating in this Nobel Symposium are not quoted, it does not imply ignorance and disregard, and I hope to get tolerant understanding for my poor and rough selection.

Kashiwabara, H., Japan. J. Appl. Phys., *3*, 384 (1964).
Nara, S., Shimada, H., Kashiwabara, H. & Sohma, J., J. Polym. Sci. A-2, *6*, 1435 (1968).
Kakizaki, M. & Hideshima, T., Repts. Prog. Polym. Phys. Japan, *11*, 351 (1968).
Nagamura, T., Kusumoto, N. & Takayanagi, M., Int. Conf. Mech. Behaviour of Materials, Abstr. P., 542 (1971).
Hoffman, J. D., Williams, G. & Passaglia, E., J. Polym. Sci. C, *14*, 173 (1966).
Nara, S., Kashiwabara, H. & Sohma, J., J. Polym. Sci. A-2, *5*, 929 (1967).
Tsuchihashi, N., Shimada, S., Kashiwabara, H. & Sohma, J., Repts. Prog. Polym. Phys. Japan, *12*, 461 (1969).
Shimada, S., Kashiwabara, H. & Sohma, J., J. Polym. Sci. A-2, *8*, 1291 (1970*a*).
Shimada, S., Kashiwabara, H. & Sohma, J., Repts. Prog. Polym. Phys. Japan, *13*, 475 (1970*b*).
Shimada, S., Kashiwabara, H. & Sohma, J., Repts. Prog. Polym. Phys. Japan, *12*, 465 (1969).
Longuett-Higgins, H. C. & Pople, J. A., Proc. Phys. Soc. A, *68*, 591 (1955).
Ohnishi, S., Sugimoto, S. & Nitta, I., J. Chem. Phys., *39*, 2647 (1963).
Shimada, S., Kashiwabara, H. & Sohma, J., Repts. Prog. Polym. Phys. Japan, *14*, 547 (1971).

Discussion

Peterlin

In order to obtain the plateaus at T_A, T_L and T_B (Fig. 1) one needs three different kinds of radicals. Each kind gets recombined according to

$$R(t) = R(0) \exp(-t/t_{dif})$$

where t_{dif} is the average life time of radicals of that kind. According to your model the life time is the average migration time of the radical to the amorphous component of the sample where the local chain mobility is sufficient for recombination. If one waits long enough ($t \gg t_{dif}$) all the radicals belonging to the plateau under consideration have recombined and hence disappeared. Since the average migration time decreases with increasing temperature one obtains at fixed time of annealing at the low temperature with a smaller than equilibrium decrease of radical population because the recombination could not be completed during annealing. By choosing a longer annealing time, the plateau would be extended in the direction of lower temperatures.

If recombination occurs only between radicals on the same molecule the average distance l they have to move is inversely proportional to the dose D

$$l = A/D \tag{a}$$

$$t_{dif} \approx l^2/r \approx 1/rD^2$$

where r is the rate of radical migration along the chain. This rate depends on temperature but not on the dose.

If recombination can occur between any two chains in the amorphous region the radicals have to move a distance about one quarter of crystal thickness L

$$l = L/4$$

$$t_{dif} \sim l^2/r \sim L^2/r \tag{b}$$

L may vary between 100 Å in single crystals precipitated from solution and 1 000 Å in bulk polyethylene annealed close to the melting point.

In the former case the crosslinking occurs on the same molecule and in the latter case between different molecules creating loops in the macromolecule and a network, respectively. The influence of molecular weight may to some extent modify the above conclusions, (case a).

My questions are: (1) Do you have any idea about the different kinds of radicals which could explain the steps in the radical decay curve? (2) Did you observe any effect of this type (dependence on D or L)?

Sohma in answer to Peterlin

(1) We concluded from the analysis of the ESR spectra that the radicals decaying in each of the three stages are not different in the species but really

different in the steric configurations (Nara, Shimada, Kashiwabara & Sohma, J. Polym. Sci. A-2, 1968). That is, the radicals decaying at the lower temperatures, T_A region, has a conformation which is less distorted from the stable form of the chain, the planar zig-zag. On the other hand, the conformation of the radicals decaying at the higher temperatures are more deviated from the planar zig-zag. In our model the radicals decaying in the different temperature regions may have different values for V_T and probably for E. In other words the less twisted radicals may migrate by the smaller thermal energy and decay in the lower temperatures. And also the migration rate may be different, presumably, depending on the radicals decaying in the different temperature regions.

(2) We did not observe the dependence of the decay rate on the dose. We determined the rate constants of the decay reaction of the radicals for each of the three stages at the roughly constant dose.

The rates (Fig. 7) reported as nearly unvaried for the different doses were not the rates of the decay but the rates of the hydrogen migration either to the adjacent site or to the sites farther separated. You have to know that the rate of the radical migration, which is same to t_{dif} in your question, is different from the rate of the radical decay. Although you assumed that the radical decays in such a way as described by the equation $R(t) = R(0) \exp(-t/t_{dif})$, this is not the case for decay. The radical decays by the recombination with the other radical, that is the decay reaction is the second order and is not described by the exponential form. In other word the radical migrates randomly until it recombines the other. That is, the migration time is closely related to the decay rate but not identical to the life time.

Thus, contrary to your anticipation the rate constant of decay, k, is independent to the radical concentration and independent to the dose. On the other hand the migration time, or the diffusion constant, D, of the migration which is same to t_{dif} in your question, may depend on the dose but not be sensitive to dose according to my model. The total number of the produced radicals is proportional to the dose but the mean distance, l, between the radicals is proportional to the cubic root of the concentration, $[R]$, that is, to the cubic root of the dose. And l is larger than the thickness of the folding crystal in the cases several Mrad. It was experimentally established that decay reaction is second order and therefore the rate, k, of the decay reaction has the dimension of $[R]^{-1}s^{-1}$. Thus, $[k \cdot R]^{-1}$ has the dimension of time, t, which means the mean life time of a radical. According to my model the mean life time of the radical t, which is equal to $[k \cdot R]^{-1}$ and proportional to the $(\text{dose})^{-1}$, is interpreted as the mean time to reach the site, like C_n, at which the radicals recombine each other. And the recombining radicals are originally separated by l. Thus, l^2 divided by this mean life time t gives the diffusion constant D because of the relation $D = \frac{1}{2}l^2/t$. Inserting the assumed dependence of both l and t on the dose the diffusion constant or the rate of the radical migration is proportional to cubic root of the dose, since

$$D \simeq l^2/t \approx (1/\text{Dose})^{2/3}/(1/\text{Dose}) = (\text{Dose})^{1/3}$$

and D depends not sensitively on the dose. This conclusion agrees with the experimental results that the migration rates are nearly constant to the dose.

Kashiwabara

In addition to Sohma's comment in answer to Peterlin's question, I would like to make the following comment: $[k\text{R}]^{-1}$ has the dimension of time as pointed out in Sohma's comment, and this is a time constant of decay reaction. The time constant of the decay reaction was compared with that of mechanical relaxation processes, and the time constant of the decay reaction was found usually to be much longer than the time constant of the mechanical relaxation. This was found in cases both of polyethylene and polypropylene. Therefore, the difference between the time constants should be corresponding to the delay due to the migration along the chain. In this respect, the situation seems to be qualitatively consistent though no quantitative estimation is established.

Sohma

As one of the collaborators of the works presented by Prof. Kashiwabara I should like to make a comment on the model for the radical migration in the polymer matrices. The majority of the radicals produced by γ irradiation in the crystalline polymers, polyethylene, polypropylene, had been primarily trapped in the crystalline part (Shimada, Kashiwabara, Sohma, Jap. J. Appl. Phys., 8 (1969)). It was found by our experiments that the decay reactions of the radicals are closely related to the molecular motions of the polymers in the amorphous parts, as mentioned in the present talk. The difference between the sites of the primary formation of the radical and the sites for the decay requires migration of the radicals in the polymer matrices. The conversion of the radical species, which had been observed in our laboratory and was mentioned by Prof. Kashiwabara, convinces us of the radical migration along
the chain. The scission radical, $\cdot\overset{\text{H}\ \text{H}}{\underset{\text{H}\ \text{H}}{\text{C--C}}}\sim$, which was mechanically produced,

was found to convert to the end radical, $\text{H--}\overset{\text{H}}{\underset{\text{H}}{\text{C--}}}\overset{\text{H}}{\underset{\text{H}}{\text{C--C}}}\sim$, which was formed by
the hydrogen abstraction not from an adjacent molecule but from the adjacent site of the same molecule. This is one of the examples of the hydrogen migration along the same polymer molecule.

Now let me present a model for the radical migration. Suppose an alkyl

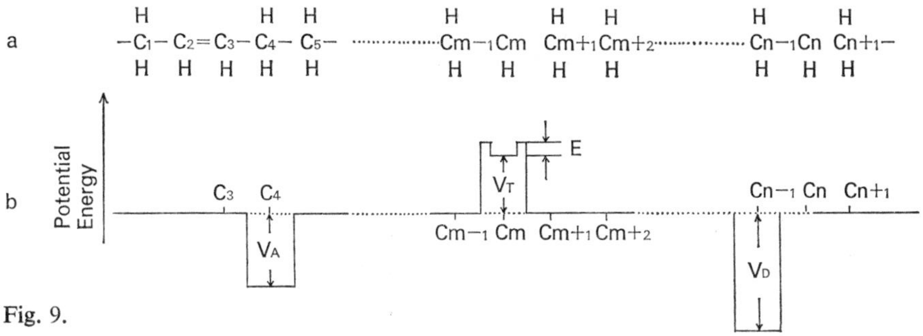

Fig. 9.

type radical of polyethylene, the so-called chain radical, $-C-C-\dot{C}_1-C_2-C-$.

Suppose the unpaired electron at C_1 abstracts a hydrogen at the adjacent site, C_2, of the same molecule and moves along the chain by one C–C unit. In this reaction of the hydrogen abstraction the final state is identical to the initial state provided with the sufficient length of the polymer. Thus, the potential seen by the unpaired electron must be flat along the chain. However, there is one thing we should not forget. The electronic configuration of the carbon atom C_1 having the unpaired electron is changed from sp^3 to sp^2, that is from the tetragonal to the planar and p_π. Thus, the misfit of the steric configuration occurs at the site of the unpaired electron. This misfit may be adjusted by a kind of internal rotation for a low molecular compound, but it can hardly be adjusted and produces the twisting of the polymer chain at this particular site of the unpaired electron at the lower temperatures like 77 K. Because the molecular motion of the polymer chain is strongly hindered at the lower temperature than those of the γ or β transition. Therefore the elastic energy is stored at the site of the unpaired electron due to the twisting of the chain. The energy at the site of the unpaired electron must be higher by the elastic energy than the other sites of the chain. This energy is expressed by V_T (twisting energy) in Fig. 9a. The unpaired electron is rather stable in the lower temperature and the radicals have a finite life time. For this stabilization the potential barrier is required and this barrier, which is observed as the activation energy for the migration, is expressed by E in the same figure. If the unpaired electron migrates along the chain, this migration is always accompanied with release of the elastic energy, that is the relaxation of the elastic energy. Suppose a site, C_4, adjacent to the double bond in the model represented as Fig. 9b. The energy for the unpaired electron at the site C_4 is lower than the ordinary sites along the chain, for the unpaired electron at this site forms an allylic radical which is more stable than the alkyl due to the resonance structure. The depth of the energy trap at C_4 is so shallow that the unpaired electron can be knocked out by UV from this trap, but the electron may be retrapped after the migration caused by thermal motion at the elevated temperatures. This is the allyl-alkyl conversion, which was observed for the polyethylene radical and also gives us the explanation why the end radical finally converts into the allyl in our experiments. Suppose the other unpaired electron exists at the site, C_n, of the same chain. The adjacent site, C_{n-1}, forms a deeper trap for the unpaired electron, because the unpaired electron at this site forms a double bond. This is one mechanism of the decay reactions of the radicals, which were found to be second order. It is, of course, probable that the unpaired electron meets the other unpaired electron belonging to the other molecule during the migration of the two unpaired electrons. The probability of such encounters must be enhanced more in the amorphous than the crystalline part at the higher temperatures than either the γ or β transition, because the molecular motions of the polymers are liberated in the amorphous region at these temperatures. This is the reason why we observe the relation between the radical decays to the molecular motions in the amorphous part, I believe.

The radical migration mechanism is nothing different from the hydrogen abstraction in the chemist's jargon and you may ask me why I assume the hydrogen abstraction only from the same chain. For the answer to this question I should like to mention two points: One is that this model is for the crystalline polymer. The nearest hydrogen to the unpaired electron on the polymer in the crystalline part is not the nearest hydrogen of the adjacent molecule but the adjacent hydrogen of the same molecule. The second is the twisting of the chain. The hydrogen abstraction leaves the twisting of the chain from which the hydrogen is abstracted. Thus hydrogen abstraction is always accompanied by the propagation of the elastic energy. Such propagation of the elastic energy may easily occur along the same chain but hardly occur between the different chains weakly coupled to each other, especially at lower temperatures than the γ transition. I do not deny the possibility of hydrogen abstraction from adjacent but different molecules at elevated temperatures at which the liberated molecular motions of the polymers permit various configurations of the chain. However, at lower temperatures the hydrogen abstraction from the same molecule, i.e. the radical migration along the chain, may prevail over the intermolecular mechanism.

Hedvig

I still wonder why this migration cannot be interpreted in terms of a random hydrogen abstraction. It is true that the initial and final stages of the reaction are energetically the same, but that of the intermediate stage is not. Correspondingly there must be a potential barrier between the consecutive states, I do not think that the barrier is only due to the chain-twisting effect.

Studies of Intramolecular Collisions between the End Groups of a Hydrocarbon Chain by ESR Technique

By Michael Szwarc

SUNY Polymer Research Center at the College of Forestry, Syracuse, N. Y. 13210, USA

On one of those long, dark, Swedish wintry nights, Professor Claesson and I talked about these marvellous polymer molecules endlessly wiggling in solution. The dynamics of their motion interested us and we were discussing various possible ways of providing information about this problem. Many have been explored, e.g., those concerned with the degree of depolarization of fluorescence emitted from dyes attached to a polymer (Frey, Wahl & Benoit, 1964; Biddle, 1968; Claesson & Odani, 1970), or variation of the polymer's NMR spectra resulting from internal rotations (Andersson, Lin & Ullman, 1970).

Then it occurred to us that the flexibility and mobility of a polymer chain may be conveniently gauged by the frequency of *intra*molecular collisions between its ends determined as a function of chainlength. How can this information be obtained experimentally? The answer is simple. Let us attach two sensors to the ends of a chain, link them to a bell that rings whenever the sensors collide and count the frequency of its strikes. It remained only to decide what kinds of sensors and bell to use and how to count its strikes. The following approach was adopted in the present investigation.

Let us link through a chain of j CH_2 units two α-naphthyl groups (α-N) which would act as the sensors. Therefore we synthesized hydrocarbons (Caluwe, Shimada & Szwarc, 1973) having the structure

$(\alpha\text{-N})\text{-}(CH_2)_j\text{-}(\alpha\text{-N})$

with $j=3$, 4, 5, 6, 8, 10 and 12. Short contact of their solution with metallic potassium reduces about 10% of them to the radical anions,

$(\alpha\text{-N}^{\overline{\cdot}})\text{-}(CH_2)_j\text{-}(\alpha\text{-N})$,

which undergo *intra*molecular electron transfer reaction through collisions between their end groups, i.e.,

$(\alpha\text{-N}^{\overline{\cdot}})\text{-}(CH_2)\text{-}(\alpha\text{-N}) \rightarrow (\alpha\text{-N})\text{-}(CH_2)_j\text{-}(\alpha\text{-N}^{\overline{\cdot}})$.

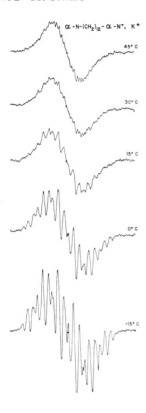

Fig. 1. ESR spectra of $(\alpha\text{-}N^{\overline{\cdot}})\text{-}(CH_2)_{12}\text{-}(\alpha\text{-}N)$ in HMPA recorded at $-15°$, $0°$, $15°$, $30°$ and $45°C$.

The contribution of *inter*molecular electron transfer to the overall process may be neglected by limiting our studies to sufficiently dilute solutions.

Provided that each collision, or rather on the average every second collision, is effective, the rate of electron transfer measures the frequency of *intra*molecular collisions between the sensors, and since such a reaction alters the ESR spectrum of the radical anions (Ward & Weissman, 1957; Zandstra & Weissman, 1962; Jones & Weissman, 1962), the shape of the ESR lines acts as our bell. Electron transfer involving free radical anions, but not their ion pairs, is essentially diffusion controlled, i.e., effective virtually on every collision. Therefore, to obtain the desired results a solvent dissociating ion pairs is required. Since hexamethylphosphorictriamide fulfills this condition (Cserheggi, Jagur-Grodzinski & Szwarc, 1969) it has been chosen as the solvent for our study.

We need now to consider the methods allowing us to deduce the frequency of *intra*molecular collisions from the shape of the ESR spectrum. However, before dealing with this problem let us see whether our sensors and our bell satisfactorily operate. Consider the ESR spectra of $(\alpha\text{-}N^{\overline{\cdot}})\text{-}(CH_2)_{12}\text{-}(\alpha\text{-}N)$ recorded at 5 temperatures ranging from $-15°C$ to $+45°C$ and depicted in Fig. 1. Inspection of this figure shows an appreciable broadening of the

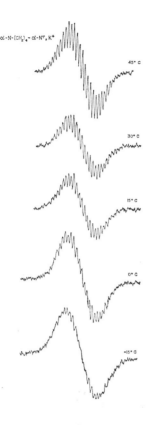

α-N-(CH₂)₄-α-N⁻, K⁺

45° C

30° C

15° C

0° C

-15° C

Fig. 2. ESR spectra of $(\alpha\text{-}N^{-})\text{-}(CH_2)_4\text{-}(\alpha\text{-}N)$ in HMPA recorded at $-15°$, $0°$, $15°$, $30°$ and $45°C$.

spectrum as the temperature rises from $-15°C^{1}$ to $+30°C$—a clear mani-festation of the expected increase in the rate of the reaction with increasing temperature. However, the spectrum recorded at 45°C is peculiar. Instead of being even broader than that recorded at 30°C it appears to be slightly sharper and it seems to reveal a new pattern of lines. One could anticipate a further development and sharpening of this pattern at still higher temperatures, because the transfer becomes then faster, but unfortunately the ensuing decomposition of radical anions prevents a direct verification of this conclusion. It is possible, however, to speed up the reaction even at lower temperatures by shortening the chain linking the naphthyl groups. Indeed, inspection of Fig. 2, depicting the spectra of $(\alpha\text{-}N^{-})\text{-}(CH_2)_4\text{-}(\alpha\text{-}N)$ recorded also at temperatures ranging from $-15°$ to $+45°C$, shows that the shape of the spectrum of this radical anion at $-15°C$ is essentially identical to that of $(\alpha\text{-}N^{-})\text{-}(CH_2)_{12}\text{-}(\alpha\text{-}N)$ recorded at $+45°C$. This implies that frequency of *intra*molecular collisions be-tween the naphthyl moieties is the same at $-15°C$ when they are separated by 4 CH_2 groups as at $+45°C$ when 12 CH_2 groups separate them.

We are able to see now how a further increase of the frequency of *intra*-molecular collisions affects the shape of the ESR spectrum. As revealed by

---

[1] It is feasible to supercool HMPA to $-15°C$ without freezing the solution.

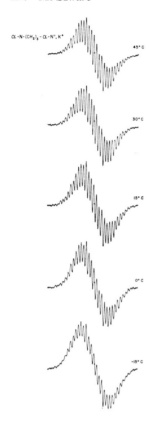

α-N-(CH₂)₃-α-N⁻, K⁺

45° C

30° C

15° C

0° C

-15° C

Fig. 3. ESR spectra of (α-N⁻)–(CH₂)₃–(α-N) in HMPA recorded at −15°, 0°, 15°, 30° and 45°C.

Fig. 2, the new pattern of lines becomes clearer at higher temperatures and quite sharp at 45°C.

The frequency of *intra*molecular collisions should be even higher for (α-N⁻)–(CH₂)₃–(α-N), and this is demonstrated by Fig. 3. The new pattern of lines, characterizing a fast rate of *intra*molecular collisions, is now quite sharp even at −15°C, and still sharper at higher temperatures. On the other hand, the frequency of *intra*molecular collisions is expected to be lower for (α-N⁻)–(CH₂)₅–(α-N) and still lower for (α-N⁻)–(CH₂)₆–(α-N). This anticipation is confirmed by the ESR spectra shown in Figs 4 and 5.

What does the new pattern of lines represent? The answer to this question calls for consideration of the problem of the so-called slow and fast electron transfer exchanges. At slow rate of exchange the ESR spectra of (α-N⁻)–(CH₂)ⱼ–(α-N) should be identical with those of a suitable model compound, e.g., *n*-butyl–α-naphthalenide (*n*-BuN⁻), when recorded in a solution of the parent hydrocarbon, *n*-butyl–α-naphthyl (*n*-BuN), of judiciously chosen concentration. In fact, had the *intra*molecular collisions in (α-N⁻)–(CH₂)ⱼ–(α-N) been prevented its ESR spectrum should be that of *n*-BuN⁻.

The above statement may appear questionable. In the *intra*molecular system

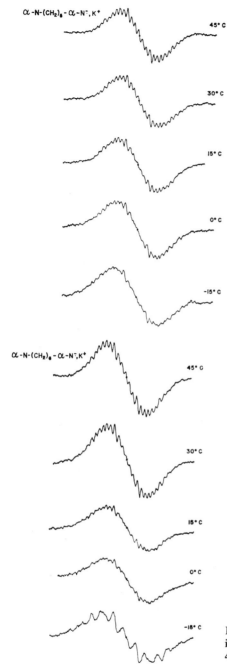

Fig. 4. ESR spectra of $(\alpha\text{-N}^{\overline{\cdot}})-(CH_2)_5-(\alpha\text{-N})$ in HMPA recorded at $-15°$, $0°$, $15°$, $30°$ and $45°C$.

Fig. 5. ESR spectra of $(\alpha\text{-N}^{\overline{\cdot}})-(CH_2)_6-(\alpha\text{-N})$ in HMPA recorded at $-15°$, $0°$, $15°$, $30°$ and $45°C$.

electron is transferred from a naphthyl moiety having a particular configuration of proton spins, let us call it the configuration *a*, to another naphthyl moiety with a spin configuration *b*. Thereafter, it is transferred back to *a*, then again to *b*, and so forth. In contradistinction, an *inter*molecular transfer,

$$n\text{-BuN}^{\overline{\cdot}} + n\text{-BuN} \rightarrow n\text{-BuN} + n\text{-BuN}^{\overline{\cdot}}$$

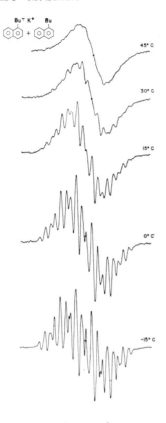

45° C

30° C

15° C

0° C

-15° C

Fig. 6. ESR spectra of n-Bu-α-N$^{\mp}$ in 2.5×10^{-2} M solution of n-Bu-α-N in HMPA recorded at $-15°$, $0°$, $15°$, $30°$ and $45°$C.

moves an electron from one naphthyl moiety to another, then still to another one, and so forth. Nevertheless, since the investigated sample of $(\alpha$-N$^{\mp})$–$(CH_2)_f$–$(\alpha$-N$)$ contains all pairs of naphthyl moieties corresponding to every possible combination of their respective spin configurations, identical spectra should be obtained for the systems $(\alpha$-N$^{\mp})$–$(CH_2)_f$–$(\alpha$-N$)$ and n-BuN$^{\mp}$ + n-BuN, provided the rates of the *intra-* and *inter*molecular transfer are equal and the transfer is relatively slow. The two spectra shown at the bottom of Figs 1 and 6 verify this point. The latter figure depicts the spectra of n-BuN$^{\mp}$ in HMPA recorded in 2.5×10^{-2} M solution of n-BuN at temperatures ranging from $-15°$ to $45°$C. The identical spectra of n-BuN$^{\mp}$ and of $(\alpha$-N$^{\mp})$–$(CH_2)_{12}$–$(\alpha$-N$)$, when both are recorded at $-15°$C, imply that at this temperature the rate of *intra*molecular transfer for $j=12$ is equal to the rate of *inter*molecular transfer, n-BuN$^{\mp}$ + n-BuN $\rightarrow$ exchange, for $[n$-BuN$]=2.5 \times 10^{-2}$ M.

However, when the ESR spectra are investigated in intermediate or fast exchange region, the results are different for the *inter-* and *intra*molecular systems, even if the rates of electron transfer are the same for both. In the limit of fast exchange the ESR spectrum of $(\alpha$-N$^{\mp})$–$(CH_2)_f$–$(\alpha$-N$)$ becomes identical with that of a hypothetic dimer $(n$-BuN$)_2^{\mp}$, whereas the ESR spectrum obtained in the *inter*molecular system loses its hyperfine structure and appears as a single, structureless line. Such a structureless line is shown at the top of Fig. 6; it represents the spectrum obtained in the *inter*molecular system at $+45°C$ when the rate of exchange seems to be very fast.

In the preceding paragraph we implied that the new pattern of lines, seen, e.g., in Fig. 3, is that of the hypothetic "dimer" $(-CH_2-\alpha-N)_2^{\bar{}}$. An electron associated with such a dimer interacts with twice as many protons as an electron located, e.g., on *n*-butyl–α-naphthalene, but the coupling constants in the former species should be $\frac{1}{2}$ of those found for the corresponding protons of *n*-BuN$^{\bar{}}$ radical anion. The ESR spectra of potassium salt of *n*-BuN$^{\bar{}}$ dissolved in tetrahydrofuran (THF) and in HMPA were recorded. The spectrum obtained in the former solvent is sharp and even the weak coupling to the 2′ protons is revealed by the partially

resolved triplet discerned in the extreme wing lines. The respective coupling constants were determined and listed in Table 1; their reliability was confirmed by computer simulation. The spectrum obtained from HMPA solution and shown in Fig. 7 is less resolved; however, its shape around its center has been found extremely sensitive even to minute changes in the values of the various coupling constants. Hence, in spite of the broadness of the spectrum, the coupling constants characterizing the radical anion in HMPA could be determined with a high degree of accuracy and the pertinent data are included in Table 1. The computer-simulated spectrum based on these data is displayed below the experimental spectrum in Fig. 7 and inspection of this figure shows excellent agreement between both. By proper choice of the linewidth we could also simulate the spectra of *n*-butyl–α-naphthalenide$^{\bar{}}$ observed in HMPA solution containing any desired concentration of the parent hydrocarbon. Again, the agreement between the computed and observed spectra is excellent, provided the comparison is limited to the slow exchange region.

Having all the required coupling constants of *n*-BuN$^{\bar{}}$ we obtained the computer-simulated ESR spectrum of the hypothetic dimer. The latter is shown in Fig. 8 and is identical with that of $(\alpha-N^{\bar{}})-(CH_2)_3-(\alpha-N)$ displayed below, proving that the new pattern of lines is indeed that of the "dimer".

We return now to the problem of determining the rate of *intra*molecular electron transfer from the shape of the relevant ESR spectrum. Different treatments are applicable to the "slowly" exchanging systems and to those exchanging rapidly.

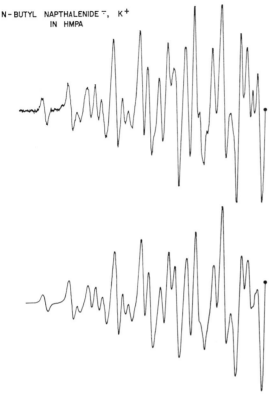

N–BUTYL NAPTHALENIDE ⁻, K⁺
IN HMPA

Fig. 7. ESR spectrum of *n*-Bu-α-naphthalenide in HMPA at room temperature (only one half is displayed). Below the computer simulated spectrum.

In the slow exchange region the spectra of the *inter-* and *intra*molecular systems may be matched and thus the frequency of *intra*molecular exchange is determined by the rate of the corresponding *inter*molecular transfer.

The rate constant of the *inter*molecular electron transfer,

$$\alpha\text{-Bu-N}^- + \alpha\text{-Bu-N} \rightarrow \alpha\text{-Bu-N} + \alpha\text{-Bu-N}^-,$$

in HMPA was determined by the method described by Weissman (Ward & Weissman, 1957). The increase in the width of the "first" line of the ESR spectrum of α-Bu–N⁻ was measured at various concentrations of the added α-Bu–N. The "first" line results from the overlap of 3 lines of a triplet with coupling constant 0.13 G (the interaction with the 2′ protons of the aliphatic chain) which are further split by the 3′ and 4′ protons (coupling constants 0.045 G and 0.035 G, respectively). The computer-drawn lines of such multiplets were obtained for various linewidths of the individual Lorentzian lines (300–900 mG), and the peak-to-peak distance of the resulting envelope was measured. Thus, the observed linewidth could be related to the "true" width of the Lorentzian lines, the resulting corrections being however small for lines

Table 1. *Coupling Constants of n-Butyl-α-naphthalenide*

$$
\begin{array}{cccc}
1' & 2' & 3' & 4' \\
CH_2\!-\!CH_2\!-\!CH_2\!-\!CH_3
\end{array}
$$

Proton	a in THF (G) at 10°C	a in HMPA (G) at 25°C	a in DME[1] (G)
2	1.60	1.53	
3	1.77	1.68	
4	4.10	4.46	
5	5.20	5.09	
6	1.65	1.62	
7	2.03	2.02	
8	5.10	4.89	
1'	2.58	2.78	2.80
2'	0.155	0.130	−0.155
3'	0.045[1]	0.045[1]	0.045
4'	0.035[1]	0.035[1]	0.035

The assignments are based on the analogy with the *a* constants reported for α-Me-naph-thalenide⁻ (R. E. Moss, N. A. Ashford, R. G. Lawler & G. K. Fraenkel, J. Chem. Phys., *51*, 1765 (1969); and α-Et-naphthalenide⁻ (G. Moshuk, H. D. Connor & M. Szwarc, J. Phys. Chem., *76*, 1734 (1972)).

[1] E. de Boer & C. MacLeen, Molec. Physics, *7*, 191 (1965). The coupling constants for 3' and 4' protons were assumed to be the same in THF and HMPA as those reported by de Boer for DME.

wider than 300 mG. A more significant correction is required to account for the overlap between the first and second line of the spectrum. Such corrections again were introduced with the aid of computer-drawn spectra.

The final calculations led to the bimolecular rate constants listed in Table 1 for the *inter*molecular electron transfer. The Arrhenius plot, shown in Fig. 9, is linear over the investigated temperature range and corresponds to the

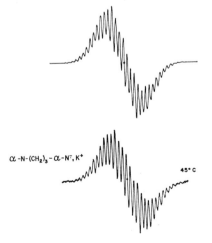

Fig. 8. Computer simulated ESR spectrum of the hypothetic dimer (–CH₂-α-N)₂⁻ (coupling constant being 1/2 of those given in Table 1); below the ESR spectrum of (α-N⁻)–(CH₂)₃–(α-N) in HMPA at 30°C.

Table 2. *Rate constants for electron transfer in HMPA, n-Bu-α-N*⁻ *+n-Bu-α-N → n-Bu-α-N +n-Bu-α-N*⁻, *for spin-spin exchange and HMPA viscosity*

T	k (electron transfer) $M^{-1} s^{-1 a,c}$	k (spin ex.) $M^{-1} s^{-1 a,b}$	η(cp)/HMPA
-15	1.9×10^8	3.4×10^8	—
0	3.5×10^8	5.6×10^8	6.23
15	6.1×10^8	7.6×10^8	4.24
30	9.1×10^8	9.0×10^8	3.03
45	11.6×10^8	14.5×10^8	2.22
60			1.71

E (electron transfer) $= 4.8$ kcal mol^{-1}
E (spin exchange) $= 3.6$ kcal mol^{-1}
$-E$ (viscosity) $= 3.9$ kcal mol^{-1}
[a] The rate constants were determined from the concentration dependence of the width of the end line of the ESR spectra. The values for the linewidth were corrected for the effects due to overlap with the next line.
[b] The spin exchange studies were performed by varying the radical concentrations from 2.63×10^{-3} M to 16.1×10^{-3} M.
[c] $[n-BuN^-] = 8 \times 10^{-4}$ M.

activation energy of 4.8 ± 1.0 kcal mol^{-1} and $A = 2.1 \times 10^{12}$ M^{-1} s^{-1}. The relatively high activation energy reflects the high activation energy of the diffusion process. The temperature dependence of the viscocity of HMPA were measured (the pertinent data are included in Table 2) and the linear plot of log η versus $1/T$ led to activation energy of 3.9 kcal mol^{-1}.

It is instructive to compare the observed rate constants with those calculated on the assumption of the diffusion controlled reaction. Assuming that equal cross sections characterize the reaction and the diffusion process, the rate constant of the diffusion controlled reaction is calculated to be 5.6×10^8 M^{-1} s^{-1} at 30°C. It is probable that the cross section for the electron transfer

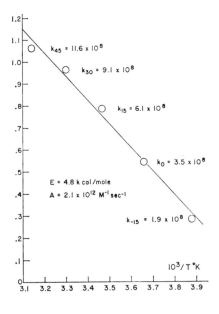

Fig. 9. *Abscissa:* $10^3/T^{-1}$/K; *ordinate:* $-8 + \log k$.

Arrhenius plot of the bimolecular rate constants of the exchange, n-Bu-α-N⁻ $+n$-Bu-α-N $\to n$-Bu-α-N⁻ $+n$-Bu-α-N $\leftarrow e$-Bu-α-N-n-Bu-α-N⁻ in HMPA.

is greater than the Stokes cross section governing the rate of diffusion and had this been the case a more realistic constant for the diffusion controlled rate might be slightly higher than quoted above. Comparison of these results with the experimentally determined value suggests that virtually every encounter is effective in the investigated electron-transfer.[1]

To strengthen this conclusion, we investigated the diffusion controlled spin-spin exchange,

$$2 \; \alpha\text{-Bu–N}^{\tau} \rightarrow \text{spin exchange}$$

The data, again derived from the broadening of the "first" line, are included in Table 2 and show that the spin–spin exchange and the electron transfer proceed with similar rates. Hence, both processes appear to be diffusion controlled.

The utilization of the matching procedure is illustrated by the following example. As was pointed out previously, the ESR spectra of $(\alpha\text{-N}^{\tau})$–$(CH_2)_{12}$–$(\alpha\text{-N})$ and of $n\text{-BuN}^{\tau}$ dissolved in 0.025 M solution of $n\text{-BuN}$ are virtually identical at $-15°C$, i.e., the rate of *intra*molecular transfer in the former system is equal to the rate of *inter*molecular transfer in the latter. The bimolecular rate constant determined at $-15°C$ is $2 \times 10^8 \; M^{-1} \; s^{-1}$, hence the frequency of the *intra*molecular collisions (twice the rate of transfer) in the $j=12$ system is $10 \times 10^6 \; s^{-1}$ at that temperature.

The *intra*molecular collisions may be treated as if they were *inter*molecular in a solution of the electron acceptor kept at the "concentration" equivalent to 1 molecule per sphere of a radius equal to the length of the extended chain. Surely, this is the *lower* limit for the more realistic "concentration" which should be still higher, because the acceptor molecule is rarely found close to the surface of the sphere—such an event requires the improbable full extension of the chain. On this basis we calculate the "concentration" of naphthyl moieties for the C_{12} system to be at least 0.07 M. However, as shown previously, the *intra*molecular transfer proceeds with the same rate as the *inter*molecular transfer when the concentration of the acceptor, i.e., $\alpha\text{-Bu–N}$, is only 0.025 M. Hence, the *inter*molecular process seems to be at least three times faster than the equivalent *intra*molecular transfer. The *intra*molecular collisions require cooperative rotation of the various C–C bonds as well as the motion of the $\alpha\text{-N}$ moieties through the viscous liquid, while only the motion of the $\alpha\text{-N}$ moieties is hindering the *inter*molecular process. Apparently the hindrance encountered in such rotations is responsible for the relative slowness of *intra*molecular collisions when compared with the *inter*molecular collisions.

The matching procedure is inadequate in a more general case of *inter*-

---

[1] To be more precise, every second encounter would be effective, since there is equal probability for the electron to remain on the original N moiety as to be transferred to the other one.

mediate or fast exchange. Our starting point in treating these cases is the solution of the Bloch equation for an absorption line arising from the electron exchange in a hypothetic molecule containing two sites of different electron-spin transition energies. The derivation based on the standard exsion gives the complex magnetization, $\hat{M}$, equal to

$$\hat{M} = \gamma_e \, H_1 \, M_o \frac{\omega - \bar{\omega} + i/T_2 + 2iP}{(\omega - \omega_A + i/T_2)(\omega - \omega_B + i/T_2) + 2iP(\omega - \bar{\omega} + i/T_2)}$$

where ω_A and ω_B are the resonance frequencies of the transition at the two sites, $\bar{\omega}$ is $\frac{1}{2}(\omega_A + \omega_B)$, and P is the frequency of jumping. The imaginary part of $\hat{M}$ gives the intensity of the ESR absorption, v, at frequency ω, i.e.,

$$v = \gamma_e \, H_1 \, M_o \{ \tfrac{1}{2} P(\omega_A - \omega_B)^2 + 2(\omega - \bar{\omega})^2/T_2 - (\omega - \omega_A)(\omega - \omega_B)/T_2 + 4P^2/T_2 +$$

$$+ 4P/T_2^2 + 1/T_2^3 \}/\{ [(\omega - \omega_A)(\omega - \omega_B) - 2P/T_2 - 1/T_2^2]^2 + 4(\omega - \bar{\omega})^2 (1/T_2 + P)^2 \}$$

Calculation of the ESR spectrum of $(\alpha\text{-}N^-)\text{-}(CH_2)_j\text{-}(\alpha\text{-}N)$ requires the addition of the v terms involving all possible pairs of proton spin configurations of the two exchanging naphthyl moieties. Due to the limitation on computer time it was not possible to consider the 2′, 3′, etc. protons in the total spin configuration. Their coupling constants are small and unresolved in the hyperfine structure resulting from the remaining protons. Hence, the 384 line spectrum of the $-CH_2-(\alpha\text{-}N)$ moiety consists of lines broader than would be expected for such a type of system on the basis of the usual T_2. This necessitated the use of a larger value for the effective $1/T_2$ in the equation giving v.

The value of the required effective $1/T_2$ was obtained by computer simulation of a line resulting from the overlap of the unresolved lines at different values of P. Such lines were computed by the method outlined above for three pairs of protons with coupling constants of 0.130, 0.045 and 0.035 G, respectively (see Table 1) assuming $1/T_2 = 0.1$ G and P equal to 1, 5, 50 and 500 G. The resulting lines are virtually Lorentzian and all correspond to an apparent $1/T_2$ of 0.2 G. Therefore this value of $1/T_2$ was used in our calculations.

The ESR spectrum of an exchanging radical anion, $(\alpha\text{-}N^-)\text{-}(CH_2)_j\text{-}(\alpha\text{-}N)$, is then calculated by adding the v terms for $(512)^2$ pairs corresponding to all the lines resulting from the electron coupling to 7 aromatic and 2 aliphatic $(C^{1\prime})$ protons. To avoid the weighing factors in the summation the latter two protons were counted as distinct although they correspond to the same coupling constant. This procedure gives the absorption spectrum and its derivative produces the conventional ESR spectrum.

The spectra calculated for P equal to 200, 50 and 5 G, respectively, with $1/T_2$ 0.2 G are shown in Fig. 10 (conversion from G to frequency in s^{-1} requires multiplication by a factor of 1.7×10^7). As expected, the computed

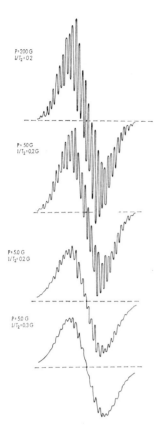

P=200 G
1/T₂=0.2

P=50 G
1/T₂=0.2 G

P=5.0 G
1/T₂=0.2 G

P=5.0 G
1/T₂=0.3 G

Fig. 10. Calculated ESR spectra of the exchanging $(\alpha\text{-N}^{\mp})$–$(CH_2)_j$–$(\alpha\text{-N})$ radical anions computed for various frequencies, P, of electron transfer.

spectra are identical with those of the "dimer", each looks as if it corresponds to a different value of linewidth of the individual lines. Hence, now we are in a position to associate the shape of an experimental spectrum with a definite frequency of *intra*molecular collisions. The relation with P is reflected, e.g., by the relative height of the teeth compared with the height of the whole spectrum. This ratio provides, in fact, a sensitive parameter for calculating P.

For P equals 0.3 G, the calculated spectrum resembles those obtained at slow rate of exchange, e.g., shown in Fig. 1.

Obviously, much additional work is needed. The effect of large end groups, acting as dampers, should be investigated. Therefore, we intend to study the behaviour of chains terminated by smaller as well as by bigger groups. Further extension of these studies calls for the investigation of chains other than $(CH_2)_j$ as well as collisions proceeding in solvents of different viscosity.

I am indebted to Drs P. Caluwe, H. D. Connor, G. Moshuk and to Mr K. Shimada who performed all the work described in this paper.

The financial support of these studies by the National Science Foundation is gratefully acknowledged.

Last, but not least my thanks are due to Professor Stig Claesson, who drew my attention to the problem of *intra*molecular collisions in polymer chains.

References

Andersson, J. E., Kung-jen Liu & Ullman, R., Discussion Faraday Soc., *49*, 257 (1970).

Biddle, D., Arkiv Kemi, *29*, 553 (1968).

Claesson, S. & Odani, H., Discussion Faraday Soc., *49*, 268 (1970).

Cserhegyi, A., Jagur-Grodzinski, J. & Szwarc, M., J. Am. Chem. Soc. *91*, 1892 (1969).

Frey, M., Wahl, P. & Benoit, H., J. Chim. Phys., *61*, 1005 (1964).

Hirota, N., Carraway, R. & Schook, W., *ibid.*, *90*, 3611 (1968).

Höfelmann, K., Jagur-Grodzinski, J. & Szwarc, M., *ibid.*, *91*, 4645 (1969).

Jones, M. T. & Weissman, S. I., *ibid.*, *84*, 4269 (1962).

Ward, R. L. & Weissman, S. I., *ibid.*, *79*, 2086 (1957).

Zandstra, P. J. & Weissman, S. I., *ibid.*, *84*, 4408 (1962).

Discussion

Yoshida

Is the dissociation of ion pairs in hexamethylphosphorictriamide virtually quantitative within the investigated temperature range?

Szwarc

Yes. Studies of electric conductivity proved that the dissociation of sodium salts of various radical anions, including sodium naphthalenide, is virtually quantitative at 25°C. The degree of dissociation increases at lower temperatures, and the temperature coefficient is too small to lead to any appreciable association even at +45°C. However, had some association occurred, the rate of the electron transfer could be only slightly *decreased* and not increased.

Kinell

You stated at the beginning of your talk that the bulkiness of the sensors is a factor affecting the frequency of the *intra*molecular collisions. Have you any information about the magnitude of this effect?

Szwarc

Not yet. We hope to extend these studies to other systems, e.g. $(A^-)-(CH_2)_{\overline{J}}-(A)$, where A denotes the 9-anthracenyl moiety, or $(p\text{-}NO_2 \cdot C_6H_4^-)-(CH_2)_{\overline{J}}-(p\text{-}NO_2C_6H_4)$. Thus, by varying the size of the sensor and determining the frequency of *intra*molecular collisions for each individual system we should be able to gauge the effect of the sensor's bulkiness on the rate.

Ivin

The frequency factor for the electron transfer reaction

$$n\text{-BuN}^- \pm n\text{-BuN} \rightarrow \text{exchange}$$

seems rather high ($A = 2.1 \times 10^{12}$ M^{-1} s^{-1}). What is the effective collision diameter for this process and how does this A factor compare with that reported by Weissman?

Szwarc

Let me point out, to begin with, that the value of the A factor claimed by us is not high when compared with other diffusion controlled reactions, and please note that the electron transfer process involving *free* ions is essentially diffusion controlled. A typical rate of diffusion controlled reaction is about 10^{10} M^{-1} s^{-1}. The "activation energy" is typically 2–3 kcal mol^{-1} giving, therefore, the formal A factor of about 10^{12} M^{-1} s^{-1}.

Returning to the electron transfer reaction, is there any theoretical justification to expect a high A factor in such a process? The answer is affirmative. The transition state of an electron transfer process certainly is not rigid, the donor and acceptor may rotate around some common axis and also may undergo various soft vibrations. Moreover, the dispersion of charge between the donor and acceptor reduces the degree of orientation of neighboring solvent molecules and this substantially increases the $\Delta S^{\ddagger}$. Hence, an A factor of 10^{12} M^{-1} s^{-1} is not surprising. In fact, in a recent study Hirota et al. (J. Am. Chem. Soc. *90*, 3611 (1968)) report log A as high as 11.8 for electron transfers involving loose ion pairs. In our own work (J. Am. Chem. Soc. 91, 4645 (1969)) we reported $A = 4 \times 10^{10}$ M^{-1} s^{-1} for the exchange N^{-} + N in HMPA. However, I believe now that this value is too low since it is based on the activation energy of 2.7 kcal mol^{-1} only—lower than the activation energy of viscosity (≈ 4 kcal mol^{-1}). The correction for the lines overlap was not introduced in that study and this may account for the discrepancy. Finally, the work of Chang & Johnson (J. Am. Chem. Soc. *88*, 2338 (1966)) gives the rate of exchange in THF for the free N^{-} with N as 3×10^{9} M^{-1} s^{-1} at 25°C. We do not know the activation energy, but assuming a low value of 2.8 kcal mol^{-1} we calculate $A = 0.3 \times 10^{12}$ M^{-1} s^{-1}.

After saying all this, I wish to stress that our value of E is not better than ± 1 kcal mol^{-1} and hence A may be lower by a factor of about 6 (i.e., it may be reduced to about 0.3×10^{12} M^{-1} s^{-1}).

Returning to your question about Weissman's result. It is impossible to overestimate the role of Weissman who is the real pioneer in this whole field. However, in his early work (1962) he was still unaware of the complications encountered in the ion pair systems. Hence, his early data with Zandstra may be questioned, in fact they lead to negative "activation energy".

In regard to the effective collision diameter I can give you only the results of unpublished calculations by Hoijtink who concluded that electron transfer may occur when the donor and acceptor are separated by 5–10 Å.

Lindberg

Did you calculate the excluded volume and virial coefficients from the data you presented here?

Szwarc

No. I doubt whether such calculations can be performed for the low molecular weight chains we are dealing with. After all, most of such calculations involve statistics assuming a long chain. In the best case, our direct results (frequency of collisions) will be represented by some indirect measure characterizing the *thermodynamic* and not *dynamic* behavior of a chain. Note that the excluded volume and the virial coefficients are not dynamic variables but variables describing the average conformation of a chain.

General Discussion

A general discussion was organized during the last day of the symposium. The discussion was concentrated around more general comments on the papers presented, the application of new ESR-techniques, e.g. ENDOR and ELDOR and new polymer fields, when the ESR method could be supposed to be more informative than other methods.

Professor Yoshida shortly communicated an examplification of the ELDOR-technique.

Professor J. Sohma communicated

ESR, which is similar to NMR in principle, may give good information on a molecular motion of polymers. The ESR parameters, such as g factor, the hyperfine coupling constant, are anisotropic. Thus the spectrum shows anisotropy for the system, of which molecular motion is completely hindered. On the other hand the spectrum must be isotropic when the molecular motion is rapid enough to average out the anisotropy. In this averaging process the ESR spectrum must change from the anisotropic pattern to the isotropic one. Thus, by observing the changes of the line shape of ESR spectrum one can obtain directly the information on the molecular motion. This analysis was successfully applied to study the molecular motions of the peroxy radicals of polytetrafluoroethylene (PTFE).

The radical produced by γ irradiation to PTFE was identified as the chain radical

$$-\overset{\text{F}}{\underset{\text{F}}{C}}-\overset{\text{F}}{\underset{\text{F}}{\dot{C}}}-\overset{}{\underset{\text{F}}{C}}-.$$

After this assignment the oxygen was introduced to the evacuated sample tube and the formation of the peroxy radical

$$-\overset{}{\underset{\text{F}}{C}}-\overset{\overset{\text{O}^{\cdot}}{\overset{|}{\text{O}}}}{\underset{\text{F}}{C}}-\overset{\text{F}}{\underset{\text{F}}{C}}-$$

was confirmed by the change of the ESR spectrum from the double quintet to the characteristic pattern of the peroxy radical. Fig. 1 shows the temperature variation of the line shape of the spectrum. The spectrum observed at 77 K was assigned to the pattern having three different principal values of g, as shown for the top pattern in Fig. 1, by the computer simulation. With raised temperatures the line shapes change. At the room temperature the one peak appears at the position corresponding to the average of g_1 and g_2 the other peak

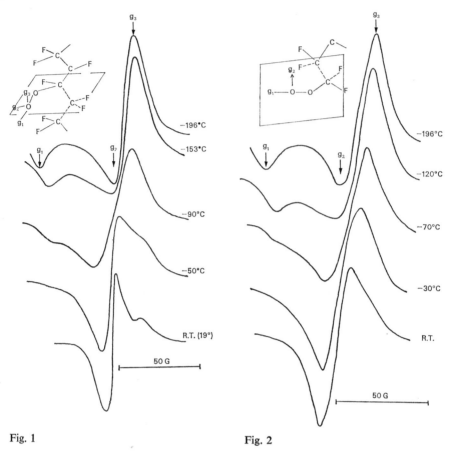

Fig. 1   Fig. 2

Fig. 1. Temperature change of the spectra observed from the peroxy chain radical. The direction of the principal axis of the g tensor is illustrated at the upper left corner.

Fig. 2. Temperature change of the spectra observed from the peroxy end radical. The direction of the principal values of g tensor is illustrated at the upper left corner.

g_3 stays almost unvaried. This indicates that the rotational motion about the axis parallel to the g_3 direction is activated gradually with the raised temperatures and the g_1 and g_2 is completely averaged out at the room temperature but g_3 is not affected by this rotational motion event at the room temperature. The direction of g_3, the smallest principal value, is known to be parallel to the molecular axis (helix axis) of the PTFE, as shown in Fig. 1. Thus the rotational random motion about the axis parallel to g_3 direction is the rotational motion about the molecular axis of the polymer.

The scission radical was formed by UV illumination to the peroxy chain radical. The formation of the scission radical was confirmed by the ESR spectrum the triple triplet and the peroxy radical was formed from this scission radical, that is ·OOC–C–. The temperature variation was traced on this

$$\begin{array}{c} \text{F F} \\ \cdot \text{OOC–C–} \\ \text{F F} \end{array}$$

peroxy radical and the result is shown in Fig. 2. The pattern observed at 77 K is quite the same as that for the peroxy chain radical. The spectrum observed at room temperature has no structure but appears as a singlet, which means complete average of the anisotropy of the g factor. This result indicates that the molecular motion of the peroxy end radical is not the rotational motion but the three dimensional random motion. The rate of this random motion at the room temperature is rapid enough to average the different principal values, g_1, g_2 and g_3.

Based on the Kneubühl's theory one can estimate the correlation time of these random motions which lead to averaging of the g factor. The averaged value of the g_1 is given by the following equation:

$$\langle g_1^2 \rangle_{\text{Ave}} = [S + (g_1^2 - S)(2/\pi)\tan^{-1}(\tau/T_2')]$$

where

$$S = \tfrac{1}{3}(g_1^2 + g_2^2 + g_3^2)$$

In these equations S was experimentally determined from the pattern observed at 77 K, which gives g_1, g_2 and g_3. The $\langle g_1^2 \rangle_{\text{Ave}}$ is the partially averaged value of the g_1 and the observed value of g_1 at the various temperatures should be used for this value. Thus, the correlation time τ of the molecular motion, which averages the g anisotropy, can be determined at various temperatures by inserting the experimental values of g_1 at the temperatures to the above equation provided with known value of the T_1^2, the line-width. The line width was determined by the computer simulation. Thus, the correlation time of τ for the molecular motion were experimentally determined at the various temperatures and the Arrhenius plots of the correlation times are reproduced in Fig. 3. The correlation times obtained from the chain radical lies on the straight line, which gives the activation energy of 0.52 kcal mol^{-1} and the correlation times for the end radical lies on the other straight line, which corresponds to the activation energy of 0.26 kcal mol^{-1}. It seems reasonable that the molecular motion of the chain radical requires the higher activation energy than for the end radical. One thing I should add is that the correlation times determined from the mechanically produced radical lies on the same straight lines to that for the end radical.

This method is being applied to the peroxy radicals of the other crystalline polymers like polyethylene in my laboratory and the results, which are in a preliminary stage at present, seem to help us in the assignment of the trapping sites to either the amorphous or the crystalline part.

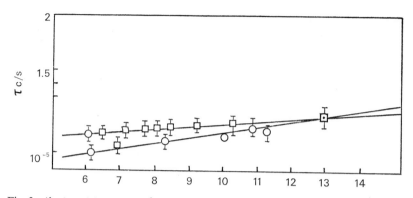

Fig. 3. *Abscissa:* 1 000 T^{-1}/K^{-1}; *ordinate:* τ_c/s.

Arrhenius plot of the correlation times. ○, the peroxy chain radical, $E = 0.52$ kcal mol^{-1}; □, the peroxy end radical produced by irradiation; △, the peroxy end radical produced by mechanical fracture, $E = 0.26$ kcal mol^{-1}.

D. R. Smith *communicated*

In solid polymers it would be of interest to measure the spatial distribution of radicals in the matrix and how it changes due to motion. This would be best done using an ESR spin-echo spectrometer such as developed at Novosibirsk. I don't believe this is commercially available yet and I am not aware of any such research outside of the USSR.

Concluding remarks

The intentions that Professor Rånby and I had when planning this Nobelsymposium No. 22 on Electron Spin Resonance applications to polymer research were to exemplify the type of information that could be obtained with this technique in various polymer fields. At the end of the symposium I feel that the contributions presented have very clearly demonstrated that during the last years ESR studies have greatly increased our knowledge not only of initiating species and polymerization reactions but also about processes going on in polymer materials under various conditions. I think that we can agree that the ESR method has given very exact and conclusive answers to questions that for many years have intrigued us. Therefore, the method has turned out to be a valuable complement to all other methods commonly in use in polymer research.

Without any doubt we can state that the ESR technique has proven to be a tool with great potentialities. However, during the lectures and discussions it has become quite evident that so far we have not used these potentialities to their full extent. Especially the discussions about rapid scan techniques contra the flow methods and about the appearance of emission lines and asymmetry in spectra prove this statement. Working along such new lines we can be quite sure that within the next few years even more detailed information about many polymer systems will be obtained. Unfortunately it was not possible to include a review paper about the ESR method as such within the frame of the symposium. It had undoubtedly been of great value to obtain information about recent trends in the development of the technique. But I am quite convinced that we all try to follow this up in order to find new approaches for solving our problems.

There is one remark that I am rather anxious to make. The presentations given have of course been concerned with fundamental phenomena. Nevertheless some papers have touched upon questions of importance for many applied polymer problems. I believe that an emphasis on such problems is quite legitimate in our present day society. The trend is to bridge the gap between science and all activities of direct importance for the welfare of mankind. As far as I can understand this is also in complete accordance with the thinking of Alfred Nobel when he decided to stimulate research by giving his prizes.

Before concluding I would like to pay some credit to Dr Don Smith from Canada who initiated the first meeting of this kind. In 1969 he arranged a small conference at the Chalk River laboratories about the use of ESR in radiation chemistry. This Nobel symposium is the second gathering concerned with the ESR technique as applied to a specified field. In my opinion meetings of this kind are valuable and necessary. Therefore I hope that somebody somewhere in the world will in the near future take the initiative to a third meeting about the application of the ESR method.

Finally it is a great pleasure for me to thank all the participants for their excellent contributions to this symposium. Professor Rånby and I are really very much obliged to all of you for making the symposium most successful. I also extend our appreciation to the Nobel Foundation for having supported this meeting in a most generous way.

Thus I declare the Nobel symposium No. 22 concluded.

Per-Olof Kinell